Dietmar P.F. Möller
Editor

Advanced Simulation in Biomedicine

With 74 Illustrations

Springer Science+Business Media, LLC

Dietmar P.F. Möller
Drägerwerk AG
D-2400 Lübeck 1
West Germany

Library of Congress Cataloging-in-Publication Data

Advanced simulation in biomedicine / Dietmar P.F. Möller, editor.
    p.   cm.—(Advances in simulation ; v. 3)
    Includes bibliographical references.
    ISBN 978-0-387-97184-1
    1. Medicine—Computer simulation. 2. Medical sciences—Computer
simulation. 3. Biological models—Computer simulation. I. Möller,
Dietmar. II. Series.
    [DNLM: 1. Computer Simulation. 2. Medicine. 3. Models,
Biological. W 26.5 A244]
    R859.7.C65A39   1989
    610'.1'13—dc20
    DNLM/DLC
    for Library of Congress   89-26082

Printed on acid-free paper.

9 8 7 6 5 4 3 2 1

ISBN 978-0-387-97184-1   ISBN 978-1-4419-8614-6 (eBook)
DOI 10.1007/978-1-4419-8614-6

Advances in Simulation

Volume 3

Series Editors:

Paul A. Luker
Bernd Schmidt

Preface

This book presents a collection of invited contributions, each reflecting an area of biomedicine in which simulation techniques have been successfully applied. Thus, it provides a state-of-the-art survey of simulation techniques in a variety of biomedical applications.

Chapter one presents the conceptual framework for advanced simulations such as parallel processing in biological systems. Chapter two focuses on structured biological modeling based on the bond graph method. This is followed by an up-to-date account of advanced simulation of a variety of sophisticated biomedical processes. The authors provide many insights into how computer simulation techniques and tools can be applied to research problems in biomedicine.

The idea for this book arose out of the daily work by experts in their field and reflects developing areas. Therefore, I think the material is timely and hope that the work described will be an encouragement for others. It is the objective of this book to present advanced simulation techniques in biomedicine and outline current research, as well as to point out open problems, in this dynamic field.

Finally, I wish to express my thanks to those colleagues who have made this book possible with their contributions.

Dietmar P.F. Möller
Mainz/Lübeck, July 1989

Contents

Contributors

Thomas G. Coleman and
William J. Gay
Department of Physiology
and Biophysics
University of Mississippi
Medical Center
Jackson, MS 39216-4505
USA

Werner Düchting
Fachbereich Elektrotechnik
Universität Siegen
Hölderlinstrasse 3
D-5900 Siegen
West Germany

E.J.H. Kerckhoffs
Faculty of Mathematics
and Informatics
Delft University of Technology
Julianalaan 132
2628 BL Delft
The Netherlands

J. Lefèvre
Department of Automatics
and Physiology
University of Louvain
Brussels
Belgium

Dietmar P.F. Möller
Physiologisches Institut
Universität Mainz
Saarstrasse 21
D-6500 Mainz
West Germany

David J. Murray-Smith
Department of Electronics and
Electrical Engineering
University of Glasgow
Glasgow G12 8QQ
United Kingdom

R.Q.Y Tham, F.J. Sasse, and
V.C. Rideout
Department of Electrical
and Computer Engineering
and Department of
Anesthesiology
University of Wisconsin
Madison, WI 53706
USA

G.C. Vansteenkiste
Department of Applied
Mathematics
and Biometrics
University of Ghent
Coupure Links 653
9000 Ghent
Belgium

Jürgen Werner
Ruhr-Universität
Institut für Physiologie
Abteilung Biokybernetik, MA 4/59
D-4630 Bochum 1
West Germany

PARALLEL PROCESSING IN BIOLOGICAL SYSTEMS

E.J.H. KERCKHOFFS
Delft University of Technology
Faculty of Mathematics and Informatics
Julianalaan 132, 2628 BL Delft, The Netherlands

G.C. VANSTEENKISTE
University of Ghent
Dept. of Applied Mathematics and Biometrics
Coupure Links 653, 9000 Ghent, Belgium

Abstract

Simulation of biological systems frequently results in models, that are very complicated and their solution on conventional computers may therefore be time-consuming.
There is a growing interest to use computers with internal parallelism and/or pipelining in this field. In this article we consider some aspects of the parallel simulation of biological systems. A general consideration on multicomputers in simulation
is presented. Simulation of biological systems and its needs for advanced computer
tools is analyzed. Finally, we give an example of biological systems simulation on
the multiprocessor AD10. The emphasis is on models characterized by ordinary differential equations.

1. MULTICOMPUTERS IN SIMULATION

1.1. Evolution_of_simulation-oriented_digital_computers

The simulation of (complex) continuous systems has been particularly influential
in the evolution of special-purpose computer systems to attain high computing speeds
and performance. In this respect two major computer architecture innovations have
permitted the circumventing of the von Neumann bottleneck in order to attain high
speeds : parallelism and pipelining. The implementation of these techniques has resulted in some distinct families of high speed digital computers including supercomputers, peripheral array processors and multiprocessors [HOC<81], [KARP84], [KARP87],
[SPRI82].

The first supercomputer to become operational (in 1975) was the ILLIAC-IV, which
consisted of an array of 64 processing elements, each containing an arithmetic/logic
unit and local memory. During the middle 1970's the focus in the development of supercomputers shifted away from arrays of processing elements towards pipelining. Two major
pipeline-oriented supercomputers, known as vector processors, of the 1970's were the
Control Data Corporation STAR-100 and the Texas Instruments ASC. The first pipeline-
oriented supercomputer to win wide-spread acceptance was the CRAY-1, developed by
Cray Research Incorporated and first installed in 1976. This machine was the trend-
setter for many other supercomputers with a similar approach to high-speed computations,
including the presently well-known products of Cray Research Incorporated and Control
Data Corporation in the USA, and Hitachi, Fujitsu and Nippon Electric Corporation in
Japan.

In the late 1970's a new class of pipeline-oriented computing devices began to

1

appear on the (signal processing and simulation) market, called "peripheral array processors" or "attached array processors" [KARP84],[LOUI81]; they are actually relatively small-scale vector processors and are intended to function as peripheral devices for sequential digital computers in order to attain high computing speed at a relatively moderate cost.

A few of the currently-available machines achieve high speed by parallelism and pipelining in more dedicated applications. An example is the Applied Dynamics International AD-10, primarily useful for simulation of dynamic systems described by ODEs [FADD84]; in this article we will present a simulation example implemented on this computer (see section 3).

A particularly important phenomenon in the evolution of simulation-oriented digital computer systems was the appearance of mini-supercomputers in the mid 1980's : commercially available multiprocessing systems and relatively inexpensive vector processors, all with a very favorable cost/performance ratio. An important subset of these mini-supercomputers shows the succesful implementation of the MIMD architecture concept, implying a (smaller or larger) network of identical processing elements (PEs).
Examples are [KARP87] : ALLIANT, FX/8   (max.   8 PEs)
                         AMETEK, SYSTEM 14   (max.   256 PEs)
                         ENCORE, MULTI-MAX   (max.   20 PEs)
                         FLEXIBLE, FLEX/32   (max.   20 PEs)
                         INTEL, iPSC   (max.   128 PEs)
                         N-CUBE, NCUBE   (max. 1024 PEs) .
Most systems employ a Unix-based operating system and are committed to providing compilers for the FORTRAN 77 and the C languages, and sometimes for PASCAL.

Another subset of the present mini-supercomputers is based upon pipelining, and is therefore more similar in organization to the CRAY and other supercomputers.
Computers of this category include : CONVEX COMPUTER CORPORATION, C-1
                                      FLOATING POINT SYSTEMS, INC., FPS 164,264 and 364
                                      SCIENTIFIC COMPUTER SYSTEMS CORPORATION, SCS-40.

1.2. Classification of multicomputers

Unfortunately, up to now no fully-satisfactory taxonomy of multiprocessors exists. It is possible, however, to identify a number of fundamental design alternatives to aid in the classification of available multiprocessors.

1) *Program control* :
   At the programming level multiprocessors can be distinguished in SIMD (Single Instruction stream/Multiple Data stream) and MIMD (Multiple Instruction stream/Multiple Data stream) architectures [FLYN72]. The key point here is whether or not the processing elements contain their own control unit.

2) *Manner of data exchange* :
   A distinction is made between shared-memory systems (possibly provided with rapid-access or cache units), in which a common memory is successively recorded and read

by the processing elements to pass information to each other, and message-passing systems. In the latter, data to be exchanged are transmitted as messages between two processing elements.

3) *Interconnection structures :*

The major distinction to be considered in network topology is whether the interconnections are dynamic or static [FENG84] . (Some authors prefer as a major division a distinction between multicomputer architectures that show or simulate a complete graph, and those that have a point-to-point topology [UHR 87]). In a dynamic interconnection network switches are provided to interconnect the processing elements (task shuffle exchange, cross-bar switch). In static networks the processing elements are permanently connected to each other in one, two or many dimensional structures. Examples of static network topologies are (see figure 1) : the star, the ring, the nearest neighbour, the systolic array and the hypercube (in a hypercube the number of processing elements, N, is made equal to a power of 2, and each processing element is connected to $2\log N$ neighbours). The processing elements are mostly communicating via a bus connection, employing such techniques as token passing and time-sliced broadcasting.

The rapid increase in the potential size of multicomputers that can be built makes feasible a great variety of possible architectures (along with a formidable set of difficult problems, that must be solved to use them effectively). Potentially, a multicomputer can be built using any possible graph topology.

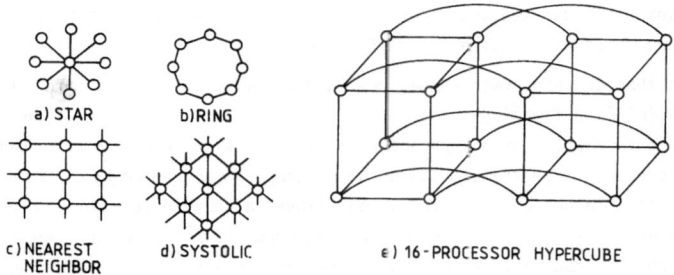

a) STAR b)RING c) NEAREST NEIGHBOR d) SYSTOLIC e) 16-PROCESSOR HYPERCUBE

Fig. 1. Examples of static network topologies

4) *Parallelism granularity :*

On the basis of the number of processing elements multiprocessors are said to be "small" (16 or less processing elements), "large" (from 16 to 1024 processing elements), or "massive" (more than 1024 processing elements).

5) *System software and languages supported :*

In the implementation of simulation models on multiprocessors the availability

3

and quality of partitioning-segmentation-synchronization supporting software play
an emerging role. Since many of the early simulation programs were written in
FORTRAN, multiprocessor systems with efficient FORTRAN compilers are at the advantage. Moreover, there is an increasing demand for compilers for more modern languages such as PASCAL, ADA and C ; also, other dedicated high level languages
(for instance, continuous simulation languages) have been proposed and sometimes
implemented [MAKO83],[FADD84] .

1.3. Parallel simulation

With the term "parallel simulation" is meant simulation on a multicomputer system.
Users of scientific digital computers are motivated to demand high processing speeds
for two distinct reasons : the increasingly detailed representations required for more
and more complex distributed parameter systems, characterized by partial differential
equations (PDEs) and real-time (or faster than real-time) computation of complex
lumped parameter systems described by ordinary differential equations (ODEs).

In the early 1980's there was talk of a centralization of supercomputer facility
(such as the Cyber 205 and CRAY-1), used among others for challenging complex simulations of systems characterized by PDEs, and accessed remotely by scientists and engineers with large simulation problems to solve. The advent of mini-supercomputers
may affect a fundamental change in this picture, since presently these are generally
marketed at a cost less than 10% of that of supercomputers (and therefore may well be
within the reach of many research groups and institutions, and may open the possibility of owning a dedicated and decentralized computer, which seems to permit a more
attractive mode of interactive scientific computing).

Rather than the performance measures in MIPS (million instructions per second)
and MFLOPS (million floating-point operations per second) provided by the vendors,
realistic performance comparisons and predictions are to be based more reliably upon
benchmark problems. For the simulation of systems characterized by PDEs the treatment
of sufficiently large sets of simultaneous linear algebraic equations (preferably with
sparse characterizing matrices) on different computers may give some performance indi-
cations ; recent studies (however, for dense matrices) showed some mini-supercomputers
to reach 5% - 20% of the performance of supercomputers, resulting in a favorable cost/
performance ratio [DONG85]. No comprehensive benchmark studies have been as yet published dealing with solving ODEs on mini-supercomputers. A general remark is that,
based on the special characteristics of the mini-supercomputer and the size of the
simulation problem (the number of state equations, the number of finite difference or
finite element points, the types of nonlinearities to be included), the user must be
enabled to provide various directives to allow the compilers to recognize paralleliz-
able or vectorizable code in order to take maximal advantage of the mini-supercomputers'
potentialities.

In addition to the afore-mentioned reasons to use multicomputers in (real-time) simulations of complex distributed or lumped parameter systems, there can be other motivations to parallel simulation. In methodology-based interactive simulation the specific architecture of such multicomputer systems might be well exploited. Focussing the attention on MIMD arrays of processing elements, examples of this in model implementation and experimentation are [KERC85],[KERC86] :

- the exploitation of the one-to-one analogy between a model structure and its physical implementation on the multicomputer ;
- model composition by assembling components running in different processing elements (configuring excitable units) ;
- interactive experimentation on model bases (for instance, multimodel output analysis after one single run).

2. SIMULATION OF BIOLOGICAL SYSTEMS AND ITS NEED FOR PARALLEL PROCESSING

Biological system modelling is influencing more and more the architectural design of simulation tools. The currently available simulators provide not enough transparency for the user who wants to model biological problems. There is, in other words, too much concentration at the moment on the mechanics of building a sophisticated computer simulation model and not enough on ensuring that the 'sophisticated' model has been adequately identified, estimated and validated.

Although almost unlimited raw computing power became available, eventually shaped in specialized tools, one must be able to construct models using entirely the language and thought patterns of the specialized field under study. Enormous benefits both qualitatively and quantitatively result when interactive tools are made available. Computer-based graphics and statistics as well as data base management and networking should be specifically designed supporting the simulationist. The interactive environment fundamentally changes the way the data are perceived and the way the modeller behaves during problem analysis and model design.

Human behaviour is strongly influenced by an interactive system's response time. According to [DOHE78] it has been observed that each second of computer response degradation leads to a similar degradation added to the user's time for the next request. This phenomenon is related to an individual's attential span. People seem to have a sequence of actions in mind, contained in a short-term memory buffer. Increases in computer response time seem to disrupt their thought processes and this may result in a time consuming rethinking of what is to be done next. In addition they become emotionally upset and make more mistakes.

The complexity of the computer model is dynamic in the sense that its content is continuously updated the more knowledge and data about the system under study are detected. [DELA83]. In order not to degrade computer response time by increasing the model complexity, the processor architecture for advanced simulators should be capable

5

to perform parallel information processing.

As indicated in detail in [SPRI82] an experimental study by simulation requires the creation and testing of an extensive variety of candidate model structures. As a consequence a simulator for biological systems must be able not only to implement and update model-banks, but also to utilize the data-banks for modelling in a methodology supported way. As a consequence simulation methodology should provide algorithms among others for :
- selection from the class of models one or more candidates, having features relevant to a certain experiment ;
- validation of a candidate model according to the particular experiment (or vice-versa) when the candidate model did not belong to the considered class.

Modelling and validation take place on experimental data and general knowledge from the actual system collected and stored in the model- and data-banks. After the validation process the experimenter may decide to continue the simulation study by further execution of simulation experiments eventually followed by another modelling phase. As a consequence a simulation study may lead to the necessity of performing modified experiments on the real process in order to get better or more relevant experimental data. In this context parallel computation is necessary to cope in an interactive way with the extensive variety of complex candidate model structures.

3. SIMULATION EXAMPLE ON THE AD10 MULTICOMPUTER

This is a report on some experiences with the Applied Dynamics AD10 peripheral processor for solving partial differential equations. PDEs cover a very wide area of applications. The present implementation was limited to the solution of equations in one space and time variable as encountered in the research on biochemical and micro-biological processes - more in particular in fermentation studies and water resources problems.

It is an obvious step trying to use the AD10 for solving PDEs, as the AD10 is especially designed for the solution of ordinary differential equations. Most common methods used for PDEs are based on the solution of systems of algebraic equations, obtained after discretization of space and time variables or on finite element techniques.
Considering the available AD10 hard- and software, we see that a semi-discretization of the problem, leaving the time variable quasi-continuous, is sufficient.

Some problems concerning wave propagation, groundwater movement and a so-called stochastic equation resulting from a stochastic approach to a microbial growth model have been solved. Results on this last application will be given next.

The equation discussed is known as the Fokker-Planck equation and is for instance used to describe one-dimensional vertical water movement in non-swelling soil :

6

$$\frac{\partial \theta}{\partial t} = - \frac{K(\theta)}{\partial z} + \frac{\partial}{\partial z} \left(D(\theta) \cdot \frac{\partial \theta}{\partial z} \right)$$

with θ the volumetric water content. The general form of the equation is :

$$\frac{\partial f}{\partial t} = \frac{\partial (g_1 \, f)}{\partial x} + \frac{\partial}{\partial x} \left(g_2 \frac{\partial f}{\partial x} \right) \qquad f, \; g_1 \text{ and } g_2 \text{ functions of } x \text{ and } t.$$

The form of the equation that will be discussed, emerged from adding a noise factor to the deterministic description of a growth model. Consider a growth model (e.g. the one proposed by Monod), that describes the growth of a micro-organism on a limiting substrate. After normalization of the biomass x to 100 and with some simplifications the model is described as follows :

$$\frac{dx}{dt} = h(x) , \qquad h(x) = \frac{(100-x)x}{K + 100 - x} \qquad \begin{array}{l} x \; : \; \text{biomass} \\ t \; : \; \text{time} \\ K \; : \; \text{growth parameter.} \end{array}$$

Its appearance is an S-shaped curve, saturating to $x = 100$ for $t \to \infty$.
A stochastic factor Ψ_ω was added to this model, and making abstraction of all the mathematical details [SPRI--] and simplifying the boundary conditions, the following Fokker-Planck equation was obtained :

$$\frac{\partial p(x,t)}{\partial t} = - \frac{\partial [\, h(x) \cdot p(x,t)]}{\partial x} + \frac{\Psi_\omega}{2} \cdot \frac{\partial^2 p(x,t)}{\partial x^2}$$

x : biomass, t : time, h(x) from the deterministic model
p(x,t) : probability distribution
$p(0,t) = p(100,t) = 0$
$p(x,0) = p_{ic}(x)$: initial distribution, e.g. gaussian or triangle around the initial condition of the deterministic model.

Semi-discretization leads to :

$$\frac{dp_i}{dt} = - \frac{h_{i+1} \, P_{i+1} - h_{i-1} \, P_{i-1}}{2 \, \Delta x} + \frac{\Psi_\omega}{2} \frac{P_{i+1} + P_{i-1} - 2p_i}{2(\Delta x)^2}$$

$$p_0(t) = p_n(t) = 0$$

$$p_i(0) = p_{ic}(i\Delta x)$$

For both first and second derivative a central difference approximation was used. This was implemented on the AD10 for n = 256 and resulted in a system of 255 coupled ODEs in the form :

$$\frac{dp_i}{dt} = \left(\frac{h_{i-1}}{2 \, \Delta x} + \frac{\Psi_\omega}{4(\Delta x)^2} \right) P_{i-1} - \frac{\Psi_\omega}{2(\Delta x)^2} P_i + \left(- \frac{h_{i+1}}{2 \, \Delta x} + \frac{\Psi_\omega}{4(\Delta x)^2} \right) P_{i+1}$$

The solution is shown in figure 2.

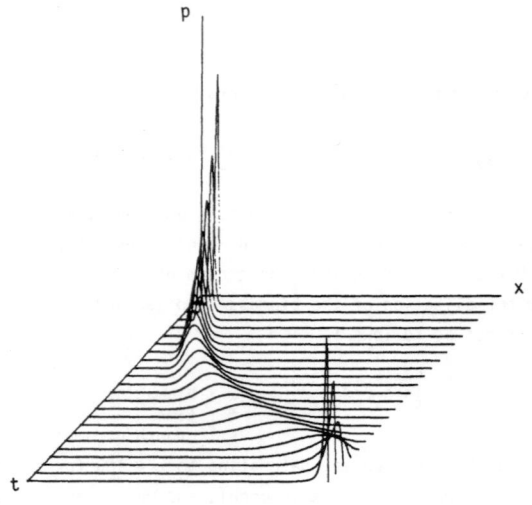

Fig. 2. AD10-solution for K = 10., ψ_ω = 1.
Total run-time = 5 seconds

The time needed to pass through all the equations (frame-time) amounts to 1,2 ms.
It was found that the solution remained stable for a Δt = 0.005 s when using the AB3
integration method, but Δt has to be less than 0.0008 for the AB1 method. A "real-time"
solution could thus be computed with a speedup factor of 4 (for the AB3 method).

4. CONCLUDING NOTES

An overview has been given how parallel computation emerges in simulation. It is shown
how the demands coming from the simulation approach to problem solution, introduces
special architectures and separate classes of mini-supercomputers. Problems dictated
by biological systems and related areas have a large impact on the further evolution
in parallel processing needs for simulation.

REFERENCES

[DELA83] DELAND, E.C. : Conceptual models in physiology, where are we ?. In : Proc.
of the IFIP Working Conference on Modelling and Data Analysis in Biotechnology and Medical Engineering, Ghent, Belgium, North-Holland Publ. Co.,1983.

[DOHE78] DOHERTY, W.J. : The commercial significance of man-computer interaction.
In : Proc. of the Conference on Man/Computer Communication, Amsterdam, 1978.

[DONG85] DONGARRA, J.J. : Performance of various computers using standard linear
equations software in a Fortran environment. Mathematics and Computer Science
Division, Technical Memorandum No. 23, Argonne National Lab., Argonne, IL.,
December 1985.

[FADD84] FADDEN, E.J. : The System 10 Plus : Broader horizons. In : W.Karplus (Ed.): Peripheral Array Processors. Simulation Series, vol.14, no.2. Simulation Councils, Inc. (Society for Computer Simulation), San Diego/California,1984.

[FENG84] FENG, T. : A survey of interconnection networks. In : K. Hwang (Ed.): Super-computers : Design and Applications. IEEE Computer Society Press, Silver Springs, MD, 1984.

[FLYN72] FLYNN, M.J. : Some computer organizations and their effectiveness. IEEE Trans-actions on Computers, vol. C-21, September 1972, pp. 948-960.

[HOCK81] HOCKNEY, R.W., C.R. JESSHOPE : Parallel Computers. Adam Hilger, Ltd., Bristol, U.K., 1981.

[KARP84] KARPLUS, W.J. (Ed.) : Peripheral Array Processors. Simulation Series, vol.14, no.2. Simulation Councils Inc. (Society for Computer Simulation), San Diego/California, 1984.

[KARP87] KARPLUS, W.J. (Ed.) : Multiprocessors and Array Processors. Simulation Series, vol.18, no.2. Simulation Council Inc. (Society for Computer Simulation), San Diego/California, 1987.

[KERC85] KERCKHOFFS, E.J.H., S.W. BROK : The Delft Parallel Processor : Properties and utilization in simulation and related fields. Systems Analysis, Modelling and Simulation (Journal of Mathematical Modelling and Simulation in Systems Ana-lysis). Akademieverlag, Berlin (GDR), vol.2 (1985), pp. 175-208.

[KERC86] KERCKHOFFS, E.J.H. : Parallel Processing and Advanced Environments in Con-tinuous Systems Simulation. Dr.-Thesis in Computer Science. University of Ghent, Belgium, june 1986.

[LOUI81] LOUIE, T. : Array Processor : A selected bibliography. Computer, sept.1981, pp. 53-57.

[MAKO83] MAKOUI, A., W.J. KARPLUS : Data flow methods for dynamic system simulation: a CSSL-IV microcomputer interface. Proceedings of the 1983 Summer Computer Simulation Conference (Vancouver, Canada). Society for Computer Simulation, San Diego/California, July 1983, pp. 376-382.

[SPRI82] SPRIET, J.A., G.C. VANSTEENKISTE : Computer-Aided Modelling and Simulation. International Lecture Notes in Computer Science, Academic Press, London 1982.

[SPRI--] SPRIET, J.A., G.C. VANSTEENKISTE : Analysis techniques for the evaluation of the representative value of stochastic models of ill-defined systems. Transactions of Simulation (in press).

[UHR 87] UHR, L. : Multi-Computer Architectures for Artificial Intelligence : Towards Fast, Robust, Parallel Systems. John Wiley & Sons, New York 1987.

AN ELEMENTARY BOND GRAPH APPROACH
TO STRUCTURED BIOLOGICAL MODELING

by J. Lefèvre

Departments of Automatics and Physiology
University of Louvain - Brussels - Belgium

1. THE NEED FOR UNIFIED GRAPHICAL STRUCTURED MODELING

In many cases, modern large biological models share the following properties:
* They are "system models" i.e. "reticulations" or interconnections of elementary
boxes representing structural and/or functional entities or basic mechanisms.
* Various submodels use different formalisms : Dependent on our goals , some parts
need physical modeling (electrical , mechanical or hydraulical networks , network
thermodynamics); others have only black box models (compartments, chemical kinetics
block diagrams.
* They are "lumped models": time is the only independent variable. In each box, the
spatial variations of internal quantities are ignored and signal exchange between
boxes is instantaneous . These models are defined by only two sets of equations:
  (i) " constitutive algebraic or ordinary differential equations " define each
        elementary box by relating constant parameters , variables and inputs ;
  (ii)" topological algebraic equations " define the constraints imposed on the
        variables by the model topology (i.e. blocks interconnection).
* They can be simulated with "state oriented simulation packages" which use an user
supplied subroutine (called here MODEL) i.e. a procedural sequence of statements
expressing the time derivatives of all state variables as functions of parameters ,
values of inputs and state variables.

When building such models , we often use graphical representations . They must ,
on one hand, summarise our ideas about the components and structural relationships;
on the other hand , they must specify unambiguously the MODEL routine. Since many
modeling methods are used, a modeling "Lingua Franca" , i.e. a unified and rigorous
formalism for all lumped systems is badly needed.

In this chapter , we survey the bases of the " Bond Graphs " graphical modeling method . In our opinion , Bond graphs (BGs) qualify as the desired "Lingua Franca" but they are often perceived as non-intuitive and hidding biological structure in a strange notation. Many modelers prefer thus less abstract diagrams (e. g. networks metabolic maps, compartments), called here "Idealized Physical Models" (IPMs) which depict more intuitively the morphology of their systems . Our goal is therefore to show that the BG-Theory may be modified to present more similarity with IPMs while retaining their simple , sound and unified theoretical basis. Due to limited space we do not give biological examples which may be found in the references.

2. GENERAL LINES OF THE PRESENTATION

BGs, defined initially by Paynter , have been developed for energetical systems in [KARN75] and extended to chemical thermodynamics in [OSTE73] . Generalisation to more phenomenological models like compartments or metabolic maps and coupling to Block Diagrams are given in [KARN81] , [LEFE85] and [DIXH77].

We start by defining hypergraphs to relate the thermodynamical ideas of [BREE81] to scattering variables ideas due to Paynter and explicited in [HOGA87] . We give next some " representation algorithms " to obtain the BGs from the usual IPMs . These algorithms are modified from [BREE86] to work in two steps :   (i) we obtain first an "isomorphic BG" (IBG) i. e. a BG showing the same topology than the usual IPM ; (ii) then , we simplify the IBG and we obtain a maximally simple BG showing energy interactions but having lost any morphological similarity with the real system We call it the "simplified bond graph" (SBG).

Until now, the intuitively meaningful IBGs have not been used by Bond Graphers who proceed directly to the abstract but computationally simpler SBGs . We claim that this fact explains the disappointment of many biologists with BGs . Computers may easily be programmed to go from one kind of BGs to the other ; both should thus be used: IBGs to think topologically and SBGs to compute and see energy flows.

Finally , we give an algorithm starting from an IBG or an SBG and writing the MODEL simulation routine . This algorithm is specially adapted to artificial intelligence programming (e. g. in "qualitative simulation" applications , see [LEFE86] . All the above papers have been very extensively used to write this chapter and the specific reference points are too numerous to be explicitly given in the following.

3. BOND GRAPH REPRESENTATION OF ENERGETICAL SYSTEMS

Electrical, mechanical or hydraulical networks, being domain dependent, cannot form the base of a unified graphical representation . To gain a more abstract point of view, we remark that , in any case, all system parts deal with energy in only a few basic ways: supply, storage, dissipation and transformation. It seems thus that we can use graphs in which nodes represent by the same domain-independent notation one and only one operation and arcs represent connections (transfer of an energy form between nodes) . However , a little thought reveals that graphs are insufficient (Fig. 1. A) . Indeed , simple arcs cannot truly to represent a power transfer between several nodes having each one and only one I/O point for energy (Fig. 1. A) : Our reticulations must at least be "hypergraphs", i. e. generalised graphs in which arcs are replaced by branches (applications between subsets of nodes (Fig. 1. B)).

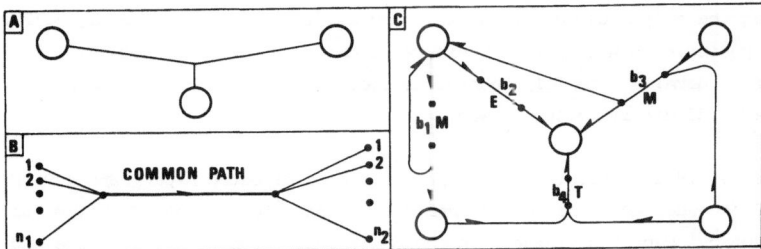

Fig. 1: A) Direct interconnection of simple nodes cannot be done by simple arcs.
   B) In hypergraphs, the complex connection of panel A is a branch joining the nodes linked to its origin arcs to thoses linked to its end arcs.
   C) An energetical reticulation with 5 nodes and 4 branches in 3 energy domains (E = Electrical, T = Thermal, M = Mechanical).

These hypergraphs represent power flows by branches defined by " branch equations " and basic operations by domain - independent nodes chosen in a set of " Elementary Component " and defined by "constitutive equations". Physical constraints on branch equations lead to the existence of two and only two kind of "elementary branches" .

By defining then "admissible branches" as a combination of "elementary branches" , we restrict our attention only to " admissible hypergraphs " having only admissible branches. Our next step is to give a graph notation for admissible hypergraphs: an elementary branch becomes an open graph with a single node called a "junction" and arcs called "Bonds" ; admissible branches become then open graphs called " junction structure graphs" and admissible hypergraphs, become "Bond Graphs".

In our presentation, hypergraphs are intermediates which can be avoided by defining BGs directly as graphs based on an extended set of nodes (elementary components and junctions). However, the definition of energy flows by a physically restricted set of branches seems much more natural . The graph notation results then naturally from these restrictions.

3.1. Hypergraph representation of energetical lumped systems

Time is the only independent variable and all the models deal mainly with energy: a real, positive, finite , time variable occuring in various forms (E_1, ... , E_n) in different domains (thermal, mechanical electrical, hydraulical, chemical.). Energy flow (P = dE/dt) is called "power".

Each model is a "reticulation" (Fig. 1. C) i.e. a directed hypergraph «EP, ET» with a set EP of nodes and a set ET of branches . The elements of EP belong to a set of "Elementary Parts" defining , on one or several forms of energy , one and only one basic operation (to be defined later). Each branch "Transfers Energy" from a given subset of EP (origin nodes) to another given subset of EP (end nodes) (Fig. 1. B) and has three parts: origin arcs, common path and end arcs . The orientations of these energy branches are given by semi-arrows.

3.2. The set EP of elementary parts or nodes

Each node exchanges energy with others in m domains and in n different ways (m≤ n). It is thus connected by n origin and/or end arcs to other nodes (e. g. in Fig. 1. C , n_A = 4) . The power flow is positive in the arrow direction . The n arcs of a node may belong to k ≤ n branches (node A , k=3). Each connection of a node with an arc is called " a port " and the node is called a "n-port".

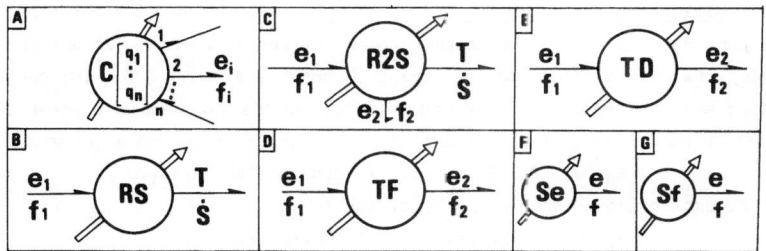

Fig. 2: BG nodes for A) capacitive n-port, B) dissipative 2-port, C) dissipative 3-
port, D) transformer (TF) or transducer (TD), E) gyrating transducer,
F) Effort source, G) Flow source. Each port may only be connected to an
input (e_i) or an output arc (e_2). Full arrows show eventual time dependance.

<u>Reversible energy Storage by n-port capacitive elements</u> : An n-port storing rever-
sibly energy in n domains is called " capacitive " and stores n thermodynamical
extensive state variables $q_1, \ldots, q_n$, the "displacements" (Fig. 2. A). The energy E is

$$E = E_q(q_1, \ldots, q_n) * E_t(t)$$

and power flow trough the ports is given by:

$$dE/dt = \Sigma(E_t(t) * \partial E_q/\partial q_i * \partial q_i/dt) + E_q * (dE_t/dt) \qquad [1]$$

The power in a port i is thus determined by two "conjugate variables"

the "flow" of q_i through port i: $f_i = dq_i/dt$ [2]

the "effort" at port i   : $e_i = \partial E/\partial q_i = E_t(t) * \partial E_q/\partial q_i$ [3]

By [3] , e_i is an eventually nonlinear "capacitive" function of $t, q_1, \ldots, q_n$:

$$e_i = Fc_i(q_1, \ldots, q_n, t) = Fc_{qi}(q_1, \ldots, q_n) * E_t(t) \qquad [4]$$

The equations [4] are the constitutive equations of the n-port. The variables q_i,
e_i and f_i used in many biological models, are given in table I. Examples of Cs may
be found in the references; if $n=1$, they are the usual capacitive 1-ports.

To be consistent with thermodynamics , each domain has a single energy form . All
the efforts are intensive and all flows are derivatives of extensities . Potential
and kinetic or electric and magnetic energies belong thus to separate domains, each
one having a single capacitive storage. These thermodynamical definitions are very
different from those used in dynamical theories (mechanics , electricity) in which
physical domains have two forms of energy (capacitive and inertial).

* Irreversible dissipation by resistive elements : The elementary resistive element
is a 2-port RS (Fig. 2. B) . Non thermal power enters at port 1 , is dissipated into
heat (Q_{cal}=S*T) and flows out at port 2 as entropy flow $\dot{S}$. The two constitutive
equations are deduced from the second principle of Thermodynamics : power cannot be
provided by the element. Thus , the first equation is an uneven algebraic relation:

$$F_d(\ e_1,\ f_1,\ T,\ t)\ =0. \qquad (d=dissipative,\ e_1*f_1<0) \qquad\qquad [5]$$

F_d may be explicited in one (or both) of the forms:

$$e_1 = F_r(e_1, f_1, T, t) \qquad , \qquad f_1 = F_g(e_1, f_1, T, t) \qquad\qquad [6]$$

The simplest R element is an electrical resistance , F_r and F_g are thus called the
resistance and conductance functions . The element must dissipate at port 2 all the
power provided at port 1. The other constitutive equation is thus:

$$e_1 * f_1 = T * \dot{S} \qquad\qquad [7]$$

T , (expressed in Kelvin) is positive ; since e_1*f_1 is also positive , the entropy
flow $\dot{S}$ =dS/dt is thus always positive and, at each time t, entropy is produced.

The usual 1-port resistive elements are obtained by neglecting the thermal port of
an RS node . Electrical resistors (Ohm's Law) , mechanical dashpots (friction law)
and hydraulical capillaries (Poiseuille's law) are constant dissipative 1-ports.
A thermal resistive 1-port (e. g. a wall) is also a dissipative element where e_1= T
and f_1= $\dot{S}$. Other examples may be found in the bibliography (unidirectional ionic
channels, cardiac valves, nonlinear viscous tissues).

The above R elements have only one non thermal port . The non thermal flow depends
thus only on one effort. However , sometimes, non thermal flow through a R element
depends irreducibly on two efforts. (e. g. radiation T1 ^{4} - T2 ^{4} or chemical
reactions , see later) . We thus define a 3-port (Fig. 2. C) with three equations:

$$f_1 = f_2 \qquad\qquad : \text{ single flow} \qquad\qquad [8a]$$
$$F_d(\ e_1,\ e_2,\ f_1,\ T,\ t)\ =0 \ : \text{ dissipation} \qquad\qquad [8b]$$
$$(e_1- e_2) * f_1 = T * \dot{S} \qquad : \text{ entropy production} \qquad\qquad [8c]$$

* Internal energy transductions by non-energic elements: Many processes transform
energy from one form into the same or another form with very small losses . These
inter-domain and intra-domain couplings , are modeled by two lossless 2-ports: the
transducer and the gyrating transducer (Fig. 2. D).

   (i) the transducer (TD) transforms non-thermal power. If both ports belong to the
same domain, it is a "transformer" (TF). It is described first by:

$$e_1 * f_1 = e_2 * f_2 \qquad\qquad [9]$$

The second equation describes the transformation of efforts:

$$e_1 = n(t) * e_2 \qquad\qquad [10]$$

n(t) is the transduction or transformation ratio (dimensionless for TFs)
From [10] and [9], we get: $\qquad\qquad$ $f_1 = (1\ /\ n(t)) * f_2$ $\qquad\qquad [11]$

(ii) the gyrating transducer (GY) is also governed by [10] but the other equations
mix efforts and flows:

$$e_1 = n(t) * f_2 \qquad\qquad [12]$$

$$f_1 = (1 / n(t)) * e_2 \qquad\qquad [13]$$

n(t) is the " gyration ratio " . We will use GYs only to unify thermodynamical and
dynamical theories. However, they have a basic role in advanced modeling.

* Power exchange with the external world by source elements : In modeling a system,
we do not study energy mechanisms used by the external world. An exchange of power
with the exterior is thus a 1-port:

$$F_s(e_1, f_1, t) = 0 \qquad\qquad [14]$$

In the first and third quadrants of the e1-f1 plane, Fs is a power supply otherwise
it is a power sink; it is usually given in one of the two forms:

$$e_1 = F_{se}(f_1, t) : \text{effort source function} \qquad\qquad [15]$$

$$f_1 = F_{sf}(e_1, t) : \text{flow source function} \qquad\qquad [16]$$

Due to dissipation , sources generate maximal effort at zero flow and maximal flow
at zero effort ; in addition they cannot supply infinite power . If we model losses
by separate R elements, ideal sources (Fig. 2. F, G) are defined by:

$$e_1 = F_{se}(t) \qquad : \text{effort source} \qquad\qquad [17]$$

$$f_1 = F_{sf}(t) \qquad : \text{flow source} \qquad\qquad [18]$$

3. 3. The two elementary connection branches

Each branch transfers power in one and only one domain (Fig. 1) and imposes on its
efforts and flows constraints extending to any domain the energetical properties of
wires and nodes in electricity or connections and rigid bars in mechanics . These
two domains are governed by the Kirchoff's laws KCL and KVL. In multienergy models
(thermodynamics , chemistry , transport) , we define "elementary branches" as those
verifying three conditions on energy: infinite speed of transport, conservation and
symmetry (i. e. independence on any ports renumbering). It is thus no more obvious
that KCL and KVL are the only valid constraints. This point will be studied now.

* Introduction of apparent scattering variables : from e and f , we define two new
variables u and v by a linear transformation:

$$u = (e/K + K*f)/2 \quad ; \quad v = (e/K - K*f)/2 \qquad\qquad [19]$$

or $\qquad e = K * (u + v) \quad ; \quad f = (1/K) * (u - v) \qquad\qquad [20]$

From [20], we get an expression for the power in terms of u and v:

$$e * f = u^2 - v^2 \qquad\qquad [21]$$

17

Usually , u and v are known as power scattering variables . This interpretation is incompatible with our infinite speed assumption, we thus introduce u and v formally from [21] and call them scattering variables just for convenience.

* The set of elementary branches: A branch has n arcs with n1 origin and n2 end arcs (n1+n2 =n) (Fig. 1). Infinite speed and energy conservation imply:

$$\Sigma_1 \, e_i . f_i = 0 \qquad\qquad \text{at all times t} \qquad\qquad [22]$$

The n arcs are also caracterized by n pairs ui,vi which by [22] , are linked by an expression of the energy conservation principle:

$$\Sigma_1 \, u_i^2 = \Sigma_1 \, v_i^2 \qquad\qquad [23]$$

The vectors $U= (u_1, . . , u_n)$ and $V= (v_1, . . , v_n)$ have thus the same length and the branch constraints are U = R. V where R is a rotation orthonormal matrix R with $R^T * R = I$.

$$
\begin{vmatrix} u_1 \\ . \\ . \\ . \\ u_n \end{vmatrix}
=
\begin{vmatrix} a\,b\,.\,.\,.\,b \\ b\,a\,b\,.\,.\,b \\ .\,.\,.\,.\,.\,. \\ b\,.\,.\,b\,a\,b \\ b\,.\,.\,.\,b\,a \end{vmatrix}
*
\begin{vmatrix} v_1 \\ . \\ . \\ . \\ v_n \end{vmatrix}
\qquad [24]
$$

By the symmetry condition, U=R. V is invariant under a permutation of i . The ortho-normality conditions are thus:

$$a^2 + (n-1) * b^2 = 1 \quad \text{and} \quad 2ab + (n-2) * b^2 = 0 \qquad\qquad [25]$$

Two trivial solutions (b=0 and a=±1) represent by [23] and [24], disconnected ports (R = I) without power exchange $e_i * f_i = 0$. The two other solutions are

$$b = \pm \,(2/n) \qquad\qquad \text{and} \qquad a = -(\pm\,(n-2)/n) \qquad\qquad [26]$$

By replacing a and b in R and expressing u_i and v_i in terms of conjugate variables, it is easy to see that these solutions correspond to two sets of conditions:

    (i) $e_1 = \ldots = e_n$ and $\Sigma_i \, f_i = 0$ \qquad\qquad\qquad\qquad [27]

   This is a "common e or KCL branch" (≈ parallel electrical connection).

    (ii) $f_1 = \ldots = f_n$ and $\Sigma_i \, e_i = 0$ \qquad\qquad\qquad\qquad [28]

   This is a " common f or KVL branch " (serial electrical connection). Equations [27] and [28] are generalised continuity and compatibility equations.

Thus, in any domain , we have only to consider these two elementary branches . For n=2, they both degenerate to the same single arc . All elementary branches are thus defined by Eqs.   [27] and [28] for any n : 2 ≤ n < ∞.   (see Fig. 3. A)

* A simple graph notation for elementary branches : We introduce a graph notation for elementary branches by defining "junctions" : two special nodes called " common effort or e-junction" and "common flow or f-junction" with respectively KCL and KVL as constitutive equations (Fig. 3. B) . An elementary branch is then replaced by its corresponding junction with n1 input and n2 output arcs. Each arc , with its single power (e, f) is called " a Bond " . For n=2 , a branch becomes an e or f-junction with one input and one output bonds and is thus equivalent to a simple bond.

18

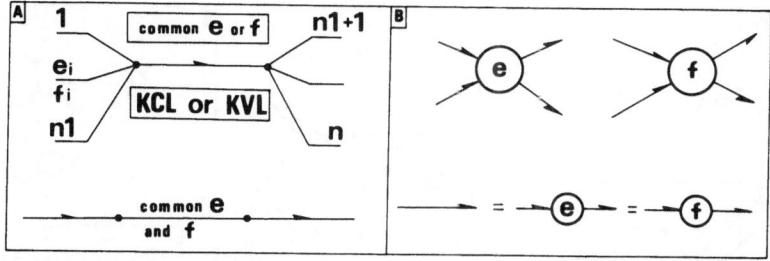

Fig. 3: A) elementary "common e or f" branches generalising parallel and serial
connections (each arc determines a power flow $e_i * f_i$). B) Graphs representing
the elementary branches of panel A.   C) Degenerate case with one input and
one output: the branch (left) and the graph become simple arcs.

3. 4. Bond graphs: a representation of a subclass of energetical hypergraphs

To "join" two branches (Fig. 4. A) , we connect one and only one output arc of one
branch to one and only one input arc of the other; the result is a more general and
" composite " branch . We may join a finite number of branches (eventually with
loops) . An " admissible branch " with n1 input arcs and n2 output arcs is then an
elementary branch or a composite branch obtained by joining an arbitrary number of
elementary branches while leaving n1 and n2 arcs free (Fig. 4. B).

A composite admissible branch keeps only two properties of the elementary branches:
i. e. infinite speed and energy conservation. An admissible energetic hypergraph is
then an hypergraph based on the set of elementary nodes and having only composite
or elementary admissible branches (Fig. 4. C) . An open graph being a graph in which
some arcs have a free extremity ; to each admissible branch, we may associate a
unique open graph by replacing each component elementary branch by its equivalent e
or f junction with its associated bonds. This graph is called "junction structure
graph" (Fig. 4. D).
Finally, a " Bond Graph " (BG) is obtained by replacing each composite branch of an
admissible energetic hypergraph by its junction structure graph (Fig. 4. E).

19

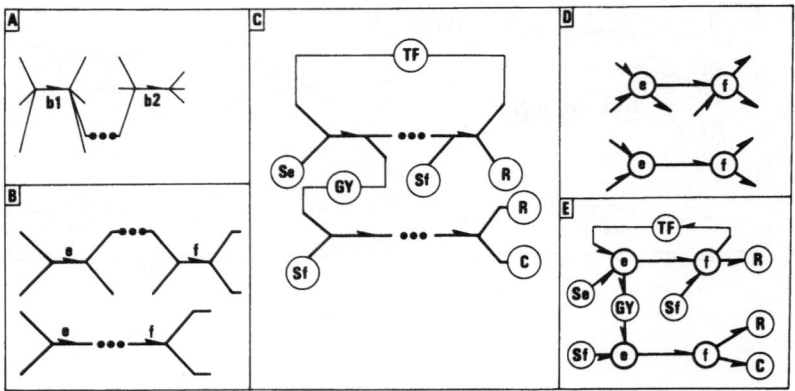

Fig. 4: A) The "join" operation for elementary branches b1=e and b2=f.

B) Two "composite branches ("join" operations are closed circles.)

C) An hypergraph connecting elementary nodes to the composite branches of panel B. Nodes and arcs orientations must match.

D) The "junction structure graphs corresponding to panel B.

E) The Bond Graph representation of the hypergraph of panel C.

4. THE CHEMICAL ENERGY DOMAIN

Like for all the other domains of Table I, we may define a thermodynamical BG-based network description of chemical reactions . Since this case is the more important from a biological point of view, we treat it in some detail.

4.1. Chemical EBG variables and energy storage elements

A solution (volume V, pressure p, temperature T, entropy S and molar numbers of the n reactants $N_1, .., N_n$) is described by a Gibbs energy storage function:

$$E = E(q_i) = E(V, S, N_1, \ldots N_n) \qquad [29]$$

μ_i being the chemical potential of species i, the Gibbs equation is:

$$dE = -p*dV + T*dS + \Sigma_i \ \mu_i*dN_i \qquad [30]$$

In the absence of reaction, the solution is thus an (n+2)-port capacitive element (Fig. 5. A). The displacements are the N_i and the flows are $\dot{N}_i = dN_i/dt$. The efforts are the μ_i. Biochemical systems being often at constant p and T ; we connect the pressure and T ports to constant effort sources . These sources are often implicit and the corresponding ports simply omitted. Correcting for their works , we get a new energy function, the Gibbs' Free energy: G= E+ p. V -T*S and the potentials are given by $\mu_i = \partial G/\partial N_i$.

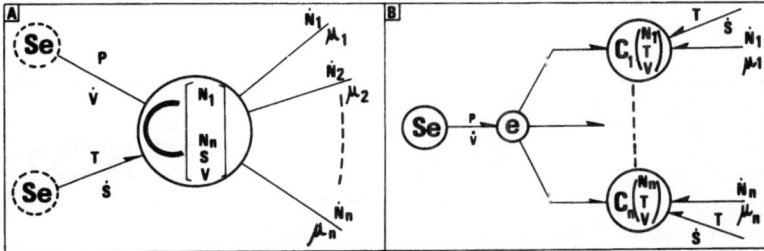

Fig. 5: A) A (n+2)-port representing a solution without reaction. Pressure and T
sources have been connected to indicate the working conditions but in
practice they are often omitted and supposed implicitly.

B) In dilute solutions, panel A degenerates in n different 3-ports; p and
T ports may be omitted in simple kinetic approaches. Here, we have
indicated constant pressure conditions but the entropic ports are free.

In concentrated solutions , each μ_i depends on all the concentrations $c_k = N_k/V$ and the n-port C cannot be simplified. However, in dilute solutions, the coupling terms $\partial^2 G/\partial N_i \partial N_j$ may be neglected . The potential is thus $\mu_i = \mu_i(N_i)$ and is frequently approximated by the familiar equation: [31]

$$\mu_i = \mu_{i,0} + R*T \ \ln(c_i)$$

The solution is then represented by n 3-port Cs (Fig. 5. B) . Each C stores a single substance and is described by the constitutive equation [31]. At the entropic port, the constitutive equation is then F(T, S, N_i)=0 and express the "heat" or "entropic capacity" of the substance.

* connection of C elements to reactions: For simplicity, we neglect p and T ports. Reactions are black box (Re) connected to the C-storages of reactants . A reactant may participate to several reactions but has only one potential; its C must thus be connected to several Re elements by an e-junction (e. g. the reactions A <----> B1 and A <----> B2 are represented by Fig. 6. A . Cs are always connected to Res through e-junctions even when they belong only to one reaction (see B1 , B2); this is the only way to preserve the orientations of bonds towards each C and through each Re.

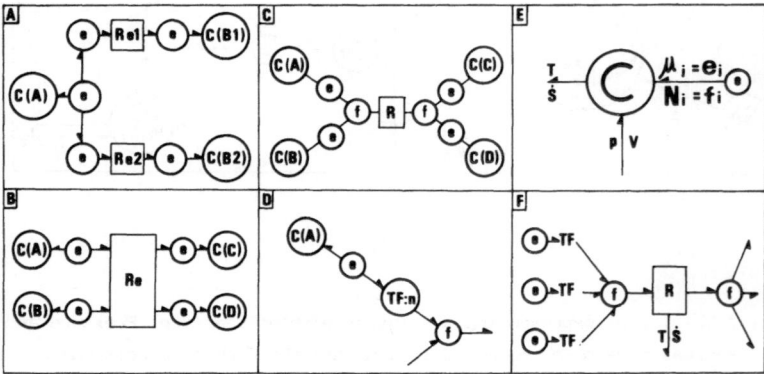

Fig. 6: A) The BG for reactions A<-->B1 and A<-->B2; C-stores are connected to Re boxes through e-junctions. B) The reaction A+B<-->C+D. C) Decomposition of the box Re of panel B. D) Representation of stoechiometry in mA+nB<-->pC+qD by transformers (only A is treated). E and F) general C and R elements.

* Internal structure of Re : the reaction A + B<---> C + D is represented by the BG of Fig. 6. B . The reaction box Re is decomposed in a 2-port R with two f-junctions. (Fig. 6. C). If the first constitutive equation of R is f1 : f2 , the flows verify the conditions fA : fB :-f1:-f2: -fC: -fD which are just those of the reaction.

To go a step further , we consider now $mA + nB <--> pC + qD$. The flow conditions are $fA/m = fB/n = -f1 = -f2 = -fC/p = -fD/q$. To get them, Re is further decomposed by introducing transformers with ratio m, n, p, q in the input bonds (Fig. 6. D).

* general C and R elements : reintroducing now T and p ports , the general diluted reactant store is given by Fig. 6. E . Its e-junction is connected to each reaction in which the reactant participates. The general Re box is given in Fig. 6. F.

* Constitutive equations of the R element: The topology of Re results only from the flow conditions and from $f1=f2$ which is one of the constitutive equations of the 3-port R of Fig. 6. F . The two other equations must be consistent with thermodynamics and with the junction structure (TFs and f-junctions). For simplicity , we redefine the stoechiometric coefficients of the bimolecular reaction as:

$$na A + nb B + nc C + nd D <------> ma A + mb B + mc C + md D$$

The new coefficients may be zero ; the left side is indicated by n and the right side by m. With $na=n$, $nb=m$, $nc=0$, $nd=0$, $ma=0$, $mb=0$, $mc=p$, $md=q$, this reaction is identical to the previous one. Introducing now global stoechiometric coefficients: $ñx = mx - nx$ for $x=a, b, c, d$; it comes: $ña=-m$, $ñb=-m$, $ñc=+p$, $ñd=+q$. The stoechiometric flow conditions of the reaction are now given by the following equalities:

$$f1 = f2 = Na /ña = Nb /ñb = Nc /ñc = Nd /ñd = fR$$

The common flow value fR is called reactional flow. The energy condition is:

$$\dot{H} = T \dot{S} + \mu a \, \dot{Na} + \mu b \, \dot{Nb} + \mu c \, \dot{Nc} + \mu d \, \dot{Nd}$$

The entropy production $\dot{S}$ is the sum of internal and external term:

$$\dot{S} = \dot{Si} + \dot{Se} \quad \text{and} \quad \dot{Se} \text{ is equal to } \dot{H}.$$

Thus: $T \, \dot{Si} + \mu a \, \dot{Na} + \mu b \, \dot{Nb} + \mu c \, \dot{Nc} + \mu d \, \dot{Nd} = 0$

Introducing the flow conditions, we get:

$$T \, \dot{Si} = (A_f - A_r). \, fR = (A_f. f_1 - A_r. f_2) \tag{32}$$

where $A_f = na \, \mu a + nb \, \mu b + nc \, \mu c + nd \, \mu d$ is the "forward affinity"   [33]

and $A_r = ma \, \mu a + mb \, \mu b + mc \, \mu c + md \, \mu d$ is the "backward affinity"   [34]

In terms of (n, m, p, q), [32] becomes

$$T \, \dot{Si} = (n \, \mu a + m \, \mu b) \, f_1 - (p \, \mu c + q \, \mu d) \, f_2$$
$$= e_1. f_1 - e_2. f_2 \tag{35}$$

This expression, consistent with the efforts $e_1 = A_f$ and $e_2 = A_r$ resulting from the BG of the reaction, is the second constitutive equation of the 3-port R.

The third equation results from the reaction kinetics : Kf and Kr being the forward and backward reaction rates , one starts from the mass action Law:

$$fR = kf * cf - kr * cr$$

with $cf = ca^{na} + cb^{nb} + cc^{nc} + cd^{nd}$

and $cr = ca^{ma} + cb^{mb} + cc^{mc} + cd^{md}$

where $cx = N_x/V$ is the concentration of $x=a, b, c, d$ To relate cf and cr to chemical potentials, we express them as follows:

$$cf = \exp((A_f - A_{f0})/(R*T)) \quad \text{and} \quad cr = \exp((A_r - A_{r0})/(R*T))$$

where we have defined the forward and backward reference affinities by:

$$A_{f0} = na\ \mu a_0 + nb\ \mu b_0 + nc\ \mu c_0 + nd\ \mu d_0$$

$$A_{r0} = ma\ \mu a_0 + mb\ \mu b_0 + mc\ \mu c_0 + md\ \mu d_0$$

replacing cf and cr in the mass action law, we get the third constitutive equation

$$fR = kf * \exp((A_f - A_{f0})/(R*T)) - kr * \exp((A_r - A_{r0})/(R*T)) \qquad [36]$$

This is the third constitutive equation of the 3-port R-element.

4.3 The Pseudo Bond graph approach to reaction kinetics and compartments

Using these BG elements , we can reticulate coupled reactions. However , a simpler "kinetic" approach is often followed . Pressure and thermal ports are neglected. Displacements are defined as "concentrations" . The efforts are simply the natural logarithms of concentrations . If [x] is the concentration of x , the equation of a C is thus $ex = \ln([x])$. R is described by $f1 = f2$ and $f1 = U(V. \exp(e1) - W. \exp(e2))$ where U , V , W are parameters . The rules to get a chemical BG are the same than before but the BG does not transport an energy ; indeed, $[x].d[x]/dt$ is not a power and is called a pseudo-power. These "pseudo bond graphs" can only be connected to a "normal BG " in a special way (section 8) . They allow nevertheless the study of enzymatic rate equations and of the models of compartmental analysis and population dynamics. Applications (enzymatic cascades , Goodwyn and Volterra networks , gene regulation, glycolysis, active transport) are described in [LEFE86].

5. REPRESENTATION OF KINETIC AND POTENTIAL ENERGIES BY DYNAMICAL BOND GRAPHS

As noted before , our BGs do not treat differently kinetic and potential energies. Both are stored in C elements belonging to separate kinetic and potential domains. In general, all kinds of interdomain energy transformations may be done by Tfs and GYs . The simplest coupling is simply a magnetic C-node (with flux = displacement, current = effort and voltage = flow) connected to a unit-ratio gyrating transducer with an electrical port 1 and a magnetic port 2. In the linear case, we have:

$$\text{for GY:} \quad e_1 = f_2 = V \quad \text{and} \quad f_1 = e_2 = I$$

$$\text{for C :} \quad \varphi = C*I \quad \text{or} \quad V = 2 = C*dI/dt$$

Eliminating V and I, we find: $e_1 = C.df_1/dt$ or $\varphi_1 = \int e_1.dt = C*I$
Renaming formally C=I, this BG is simply the electrical inductance element. In the same way , in mechanics or hydraulics , a kinetic C connected to a unit ratio Gy

between the potential and kinetic domains is equivalent to an inertial mass. By defining GY→ C as a new "elementary inertial component" I, thermodynamical and dynamical approaches are unified. This entails the definition of a new fundamental variable : the momentum p which is the derivative of the effort.

6. ALGORITHMIC REPRESENTATION OF IDEALIZED PHYSICAL MODELS BY BOND GRAPHS

The e and f junctions generalising serial and parallel electrical connections, the circuit of Fig. 7. A is equivalent to the BG of Fig. 7. B . Both are described by the sentence "SE is serially connected to R2 , C2 and the parallel connection of R1 and C1" . However , the BG has lost a good deal of the intuitive interest of the circuit. The same conclusion holds for all dynamical energy domains . Thus , for these systems , we start modeling not from BGs but from domain specific idealized physical models or IPMs (e. g. Fig. 7. A) . Since we want a unified but intuitive formalism, the problem arises to translate algorithmically IPMs into modified BGs preserving , if possible, physical intuition. We give now a "representation method" achieving this result in any domain and building a BG with the same "appearance" than the IPM. This BG may be simplified but then similarity with IPMs is lost.

6.1. The EBG representation of an electrical IPM (i. e. circuit)

The representation algorithm has eight steps illustrated on the IPM of Fig. 8.
1. Draw a set (minimal or more extensive) of nodes separating elements and/or ports (Fig. 8. A).
2. Represent (Fig. 8. B) these nodes by e-junctions (common potential points)
3. If an element or port is connected to two nodes , insert by an f-junction , the corresponding BG element or port between the two e-junctions representing these nodes (Fig. 8. B). Impose bond arrows as shown since the currents in a dipole and in its connecting wires are equal and since every dipole feels the effort difference between poles . (orientations give signs in constitutive equations)
4. Potentials having a ground reference , connect a zero e-source to the ground e-junction (Fig. 8. B)

These four steps result in a fully specified BG similar to the IPM and which may be used to write simulation routines.

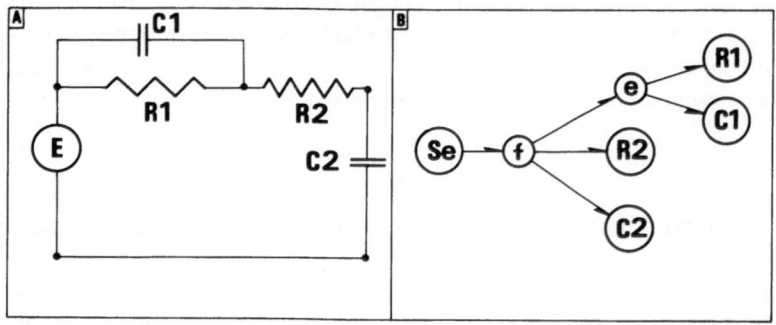

Fig. 7: A) An intuitive electrical IPM. B) the Bg equivalent to panel A.

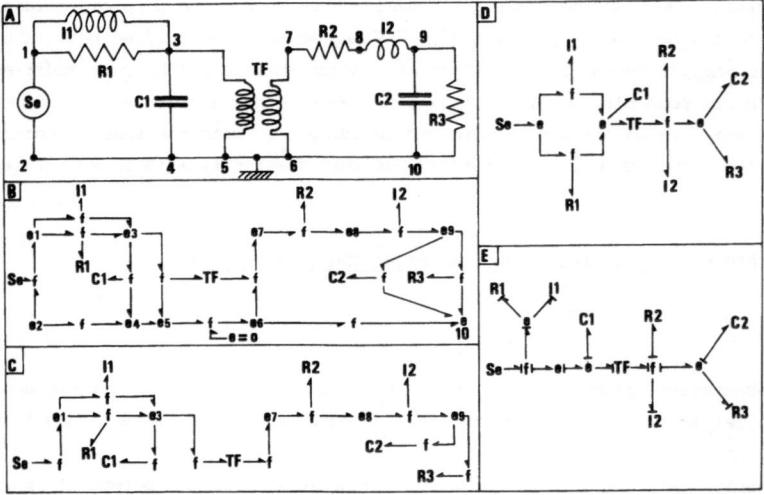

Fig. 8: The steps of the representation algorithm

We can stop the algorithm here. However, to write equations we can simplify the BG using rules described in the next four steps (Fig. 8).

5. There is no power in a bond connected to ground , delete the ground junction and all its bonds (Fig. 8. C).

6. Replace junction paths in which junctions have only two bonds by single bonds (e. g. TF, f, e7 or SE, f, e1). Redraw the BG (Fig. 8. D).

7. Replace sequences of e or f by single e or f junctions (rules in Fig. 9. A)

8. An n-sheet is a structure made of two e-junctions connected by n f-junctions (Fig. 9B). BG parts are connected by single bonds to the two e-junctions and to the n f-junctions . If the Bg has n-sheets , simplify them by using the rule of Fig. 9B . When several BGs are attached at a e-junction of a sheet (Fig. 9. C) , introduce an extra junction before simplification . The result of steps (6, 7, 8) is a maximally simple BG (Fig. 8. E) which has lost its initial "realistic" look but maximizes the simplicity of the procedure to write the simulation routine.

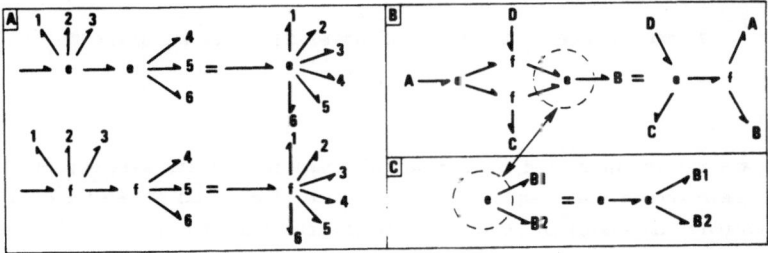

Fig. 9: Simplification rules of BGs

6. 2. Representation algorithms for other systems

Hydraulical and electrical systems have isomorphic IPMs . We can therefore obtain hydraulical BGs by the preceeding method applied to hydraulical variables . A similar method exists for 1D mechanical systems . It begins by identifying nodes of common velocity (f-junctions) separating the elements. Elements are introduced between their connecting f-junctions by an e-junction (common force) . A reference may be introduced as a source of zero velocity . The result is a BG similar to the original IPM. The steps 5, 6, 7, 8 are unchanged but simplification rules are dual.

In thermal systems, independent T s are replaced by e-junctions and the elements are introduced by f-junctions. The same simplifications as above may be applied and the simplified BG shows clearly the morphology of the system. Finally, chemical Bgs are still more intuitive. If, for simplicity, we neglect the entropic ports and start from a list of reactions, their representation algorithm has four steps:

* create a reactant store for each species Xi.
* create a reactional block Re for each reaction.
* connect the e-junctions of Cs to the reactions in which they participate.
* suppress redundant TFs (coefficients equal to 1).

The resulting chemical Bg is much more intuitive than the list of reactions since it shows all the reactional couplings, feedbacks and interactions [LEFE 85]. In biochemical systems, the BG has the same topology than the biochemical maps used so fruitfully by biologists to study complex reactional pathways.

7. COUPLING OF BOND GRAPHS AND INFORMATIONAL BLOCK DIAGRAMS

When we are mainly interested in information transfers between system parts, we may neglect their energetical coupling. For instance if a signal, taken from a system part, controls the system by acting on a parameter of another part, the power used for measurement and control is often negligible with respect to the main power. In addition, our knowledge is often so scarce that, for many parts, we cannot build energy models but only phenomenological equations (e.g. the Guyton model of body fluid regulations) describing pure information flows which cannot be directly exchanged with a BG. (e.g. pseudo BGs, compartmental models, block diagrams). We need thus models of unidirectional energy-less " pure signal " flows. The usual block diagrams (BDs) are well suited to this role and must be coupled to BGs.

7.1. Representation of information transfer in BGs by active bonds:

An energy-less, unidirectional signal transfer between two BG nodes is drawn as an "active bond" : a full arrow between the emitter and receiver nodes. Thus A-s->B represents the sending of a pure signal s from node A to node B. In the same way, Fig. 10. A represents the sending of signals (s1,..,sm) from nodes (A1,..,Am) to a block diagram BD computing signals y1,..,yk sent to nodes (B1,..,Bk). BD may be simple (analog computing) or complex (e.g. filtering, control, optimisation).

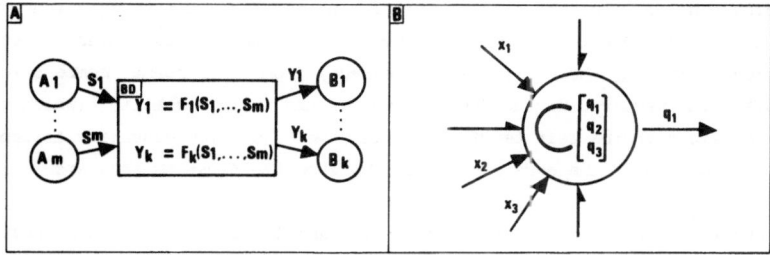

Fig. 10: A) Connection of BG parts to a block diagram
         B) Input (x1, x2, x3) and output (q1) informational ports

7.2. Coupling of active bonds to BG nodes

In BGs, "pure input signals" always modulate parameters and thus may be sent to any
elementary node but not to junctions . If an n-port receives m signals , we add to
its representation m informational input ports coming from active bonds (Fig. 10.B).
Output measurements may be displacements, momenta, efforts or flows. Displacements
and momenta exist only in Cs and Is. Their measurement in these nodes is indicated
by issuing an active bond from an output informational port added to the element.
(Fig. 10.C). Unambiguous efforts and flows exist only in e and f-junctions. So, to
measure an effort , we issue an active bond from an informational port added to the
e-node where this effort is constant. If there is no such e-node , we augment the
BG by adding a new 2-port e-node in the bond to which the effort belongs. The same
procedure applies to flows in f-nodes . With these definitions , graphical models
of energy and information flows become possible in a unified way.

8. FORMULATION ALGORITHMS: AUTOMATIC OBTENTION OF SIMULATION ROUTINES

A simulation routine (MODEL) is a procedural sequence of (eventually conditional)
assignments ; each assignment computes one of the variables of an element or of a
Kirchoff law. The obtention of MODEL is thus a four steps process:

* choose adequate functional forms for each constitutive equations.
* choose and isolate "the good " variable in each constitutive equation . This is
  called the " element causality assignment".   Indeed, the assignments are seen as
  devices computing their left member (effect) from their right members (causes).
* choose the required Kirchoff's constraints and their causality (which variable
  to explicit in the left term) . This is the " junction causality assignment" .
* sort procedurally the equations resulting from the first three steps.

The "causal or computational structure of the model", is the procedural sequence of
causes and effects determined by MODEL regardless of the exact equations . Usual
networks represent elements by arcs and connections by nodes . The two kind of
MODEL lines having thus different graph representations, the writing of MODEL needs
advanced topological methods , is domain specific and gives not much insight into
the computational and/or physical structures.
By contrast, BGs represent both elements and junctions by nodes and thus treat each
line of MODEL similarly. As a consequence , formulation of MODEL is elementary even
for nonlinear , time varying systems ; it is domain independent and allows a great
deal of insight into both computational and physical structure.
We describe first the " causal stroke", a tool introduced by Paynter to help in the
causality assignment. Then we describe the MODEL formulation method : the causality
assignment is done for both elements and junctions by a first algorithm; then paths
of causalities are followed in the BG by a second algorithm to write MODEL.

8.1. the causal stroke: a graphical marker of causality

A bond linking two subsystems A and B may have two orientations :   A——≫X or A≮—X .
In both cases , the e and f signals are , physically and computationally , the only
messengers of information between A and B in this bond. Since they must be unambiguously determined by the equations of the BG, one of three cases must occur:

* e , resulting from the equations of A , is imposed to B . Reciprocally f , coming
  from the equations of B, is imposed to A. We represent this causality by:
  $$A \longrightarrow\!\!| B \quad \text{or} \quad A \longleftarrow\!\!| B$$
  The bar is called "causal stroke" and , independently of bond orientation , points
  towards the side receiving the effort as a cause (right member of an assignment).
  Flow is causally imposed in the reverse direction.
* the value of e is imposed by B to A which in turn computes f and gives it to B:
  $$A |\!\longrightarrow B \quad \text{or} \quad A |\!\longleftarrow B$$

* the values of e and f are determined simultaneously by the equations of A and B :

$$A \longleftarrow\mkern-6mu| B \quad \text{or} \quad A \mkern-6mu|\longrightarrow B$$

This case known as a " causal conflict or implicit equations loop " indicate the presence of a modeling problem and is therefore exceptional (see later).
Two orientations caracterize thus a bond: its power orientation given by our former representation convention and its computational causal stroke . BG elements impose four kinds of constraints on their strokes i. e. on their MODEL assigments.

TYPE 1: PORTS WITH IMPOSED CAUSALITY

* a linear or controlled source imposes respectively

        the effort for an e-source: $\longleftarrow$ Se or $e=F_{se}(t)$

        the flow for an f-source : $\longleftarrow\mkern-6mu|$ Sf or $f=F_{sf}(t)$

* any non-linear element may impose the causality of some ports if some of its constitutive equations, being not invertible, admit only one form of assignment.

TYPE 2: PORTS WITH COMPUTATIONALLY PREFERABLE CAUSALITY

* a capacitive element admits in principle two causalities: the first form is the " integral causality " : q is a "state variable" imposed initially and known by integration of a flow determined outside. Conversely , e is determined from q:

$$\longmapsto C \quad , \quad q=q(0) + \int f \dot{} dt, \quad e=q/C$$

In the second case, e , imposed from outside , determines q. The flow f is then obtained by differentiation (derivative causality):

$$\longrightarrow\mkern-6mu| C \quad , \quad f=C. \, de/dt, \quad q=Ce$$

Numerical integration being superior to numerical differentiation , we exclude the derivative causality and admit only n-port Cs in integral causality . All the displacements of a BG appear thus as "state variables".
* Similarly , linear or nonlinear inertial elements will be considered only in the integral causality ($\longrightarrow\mkern-6mu|$ I , $p = p(0) + \int e dt$, $f = p/I$) and derivative causality ($\longmapsto$ I , $e = I \, df/dt$, $p = If$) will be excluded for the same reason. Thus, the momenta in the Is of dynamical BGs are also state variables.
If a BG does not allow integral causality on all Is and Cs, infinite transients may be generated. From a physical viewpoint, the most satisfactory solution is to modify the BG by adding for example parasitic losses which were neglected at first sight. However, this may create numerical stiffness problems.
* Any element having a constitutive equation computationally difficult to invert has a preferred causality to specify initially.

TYPE 3: N-PORTS WITH CAUSAL CONSTRAINTS

* Due to their equations , the transformers have only two imposed causalities : The first form is $1 \longrightarrow\mkern-6mu|$ TF $\longrightarrow\mkern-6mu|$ 2; TF receives e_1 from port 1 and imposes $e_2 = e_1/n$. The second is $1 \mkern-6mu|\longrightarrow$ TF $\mkern-6mu|\longrightarrow$ 2; the situation is opposed: $e_1 = n*e_2$ and $f_2 = n*f_1$.

* For gyrating transducers, the situation is similar:

$$1 \longrightarrow GY \longmapsto 2 \qquad f_2 = e_1/n \quad \text{and} \quad f_1 = e_2/n$$

or $\qquad \longmapsto GY \longrightarrow 2 \qquad e_2 = n \ast f_1 \quad \text{and} \quad e_1 = n \ast f_2$

* In f-junctions, we may explicit one effort in $\Sigma e_i = 0$: external elements may impose any set of (n-1) efforts on the junction but the last one is determined by KVL; f-junctions must have strokes in all bonds except one (Fig. 11. A).
* For an e-junction, the situation is dual; one of the bonds must impose e and the other efforts are imposed by the junctions (effort equality) (Fig. 11. B).

TYPE 4: PORTS WITH AN INDIFFERENT CAUSALITY

A resistive port with a theoretically and practically invertible constitutive law may be described by two causal forms:

$$\longmapsto R, \; e = R \ast f \qquad \text{or} \qquad \longrightarrow R, \; f = e/r$$

In nonlinear or time varying element, it may happen that, for some values of parameters or variables, such an element degenerates and imposes a causality.

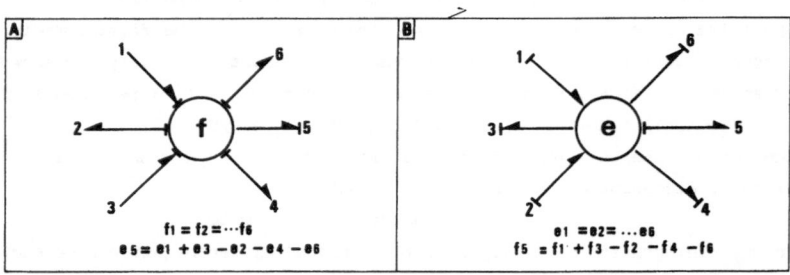

Fig. 11 Causality on f and e junctions.

8. 2. Sequential assignment of causality for elements and junctions

We will demonstrate that if each bond of a BG may receive one and only one stroke verifying the laws defined in 8.1 then all the junctions and elements equations (i.e. all the lines of MODEL) may be explicited correctly. The BG is then called a "Causal BG". Before to give a method to place such consistent causal strokes in BGs, two notions must be defined :

* It may happen that a BG imposes some causal strokes inconsistent with the rules defined above for fixed or prefered causalities and for n-ports causal constraints. In general, possible solutions to these "causal conflicts" are replacement of some

32

elements by a single one, choice of non prefered causalities or model modification. In our treatment , prefered causalities being imposed , the model must be modified.

* A "causal path" between two elements A and B is a path of junctions or n-ports joining A and B with causalities going in one way (e. g.   A⇀|e⇀|f⇀|TF⇀|B).

The sequential causality assignment is then the following:

    Assign causal strokes to all elements of fixed causalities (type 1).

  <u>repeat</u>

    Apply all the necessary causal constraints of type 3 to elements or junctions connected to bonds having already received a causal stroke.

  <u>until</u> It exists no elements for wich a type 3 constraint is satisfiable.

  <u>if</u> causal conflict occurs   <u>then</u> Change the model (it is ill defined) .   <u>endif</u>

    Assign causal strokes to each prefered causality (type 2).

  <u>repeat</u>

    Apply all the necessary causal constraints of type 3 to elements or junctions connected to bonds having already received a causal stroke.

  <u>until</u> It exist no elements for which a type 3 constraint is satisfiable.

  <u>if</u> causal conflict occurs <u>then</u>

    <u>if</u> causal paths occur between storage ports of the same nature   <u>then</u>

      Condense these elements in one element.

    <u>endif</u>

    <u>if</u> preferable causalities may be changed   <u>then</u>

      change causality of type 2 ports to suppress conflicts.

    <u>else</u>

      Change the model.

    <u>endif</u>

  <u>endif</u>

  <u>if</u> at least one port of type 4 has no stroke   <u>then</u>

    Some R-elements have free causality : it exists a causal path between them. There is an algebraic loop i. e. a system of simultaneous equations . We have four solutions:

        solve the system (algebraically or numerically at each step),

        or introduce a delay in the loop if gain smaller than 1,

        or condense some R-elements together,

        or modify the model by adding parasitic storage elements.

  <u>endif</u>

  <u>end</u>

In the final panel of Fig. 9 , the causal strokes have been placed according to this algorithm . Its application to a complete IBG is also possible but somewhat longer. Analysis of the more frequent causal conflicts may be found in the literature.

8.3. Algorithmic derivation of the procedural simulation routine:

For speed reasons, previous formulation algorithms acted on data structures needing a numbering procedure for elements, ports, junctions and bonds. In large models, these number notations become mind-boggling. In contrast, we will now define an algorithm, based on the use of mnemonic names and local causality rules, which duplicates the way humans think about their models. This algorithm is slower but the automatic obtention of MODEL becomes elementary and intuitively meaningful. In this presentation, we consider only BGs without informational ports and in which elements are always separated by junctions. (e.g. Se—TF—C will be transformed in Se-e-TF-e-C). All elements receive a name beginning by the mnemonic of the element type (Se, Sf, R, RS, C, I, TF, GY) followed by an identifier (alphanumeric string). The displacements and momenta are denoted by Q or P followed by the element name; the efforts and flows in " external bonds " i.e. those attached to an element are denoted by E or F followed by the element name and, for n-ports, by a port indix; junctions are denoted by e or f followed by a number. Finally, "internal bonds", connecting junctions together are denoted by concatenating their junctions names.

At run time, MODEL starts from input time signals (U1,..., Uk) and state variables (QC1,..., QCn and PI1,.., PIm) and ends up with assignments for state derivatives (flows in Cs: FC1,..,FCn and efforts in Is: EI1,..,EIm). Starting from a causal BG, the formulation algorithm obtains the assignments in the reverse order:

* for each state variable Xi define an empty list ASSIGNMENTS_OF_Xi.
* define an empty list CURRENT_ASSIGNMENTS.
* define a set STATEVAR containing the state variables.
* define a list STATE_TO_TREAT containing also the state variables.
* put the input signals and the Xi in the set CURRENTLY_ASSIGNED.

Repeat
    * delete the first element of STATE_TO_TREAT, put it in VAR_TO_TREAT.
    * put in UNASSIGNED the derivative of the Xi in VAR_TO_TREAT.

    Repeat

      Repeat
        * Apply procedure CAUSAL_RULES, to be defined below, to write and put in CURRENT_ASSIGNMENTS assignments for the variables of UNASSIGNED which are then transfered to CURRENTLY_ASSIGNED.

      Until UNASSIGNED empty

    * Put in UNASSIGNED the variables of the right members of all the assignments written in the preceeding step and not belonging to CURRENTLY_ASSIGNED.

Until UNASSIGNED empty.

* Put in ASSIGNMENTS_OF_Xi the assignments in CURRENT_ASSIGNMENTS.
* make VAR_TO_TREAT and CURRENT_ASSIGNMENTS empty.

Until STATE_TO_TREAT is the empty list

* concatenate all the lists ASSIGNMENTS_OF_Xi in a list MODEL.
* in the assignments of MODEL , replace all efforts and flows of internal bonds by the efforts and flows of external bonds to which they are equal.
* delete the assignments of internal efforts and flows from MODEL.
* sort MODEL procedurally.

end

It remains only to specify the procedure CAUSAL_RULES . We do it informally by an example using many of these rules to build MODEL for a simple BG (Fig. 12.) . From these examples and from the causality rules defined in section 9. 2 and 9. 3, one may easily infer the finite set of rules used in our algorithm.

This algorithm may be applied indifferently to multi-energy IBGs or BGs. In both cases, the efforts and flows in internal bonds may be eliminated from the equations which are then identical and maximally simple. Let us call " causality tree CT(x) " of a variable x , a tree in which : (i) each node is a variable of MODEL ; (ii) the root is x ; (iii) state variables and input signals have no decendants ; (iv) the direct decendants of a variable y are the variables present in the right member of the assignment of y in MODEL . In fig. 13 , we show the causality trees of the state derivatives of Fig. 12 . CT(FC1) is completely drawn , but CT(FC2) has been prunned of the descendance of FL which is already represented in CT(FC1). In the same way, CT(VL) is prunned from the descendance of VC1, FL and VC2 which occur in CT(FC1) or in CT(FC2). These trees, directly constructed during the determination of the lists of assignments of each state variable, are very useful to sort the equations either for sequential or parallel simulation algorithms . Indeed , they show directly the precedence relations between the sets of assignments for each state variable . In a sequential simulation , these sets must be sorted in the order (A , B , C). In a parallel simulation (e. g. on a bank of transputers in Occam), the set A, B, C are parallel processes and their precedence relations become the starting conditions of processes. Moreover, each branch could be implemented on a different transputer. BGs have therefore direct implications for massively parallel simulation.

These trees may also be used to guide a reasoning process to infer non - monotonic "sequences of causal qualitative or linguistic explanations on a BG behavior . In these applications, a MODEL assignment is seen as a microscopic inference (e. g. if I large and R large then V very large") and a sequence of microscopic inferences is an "explanation" which may be compiled or aggregated . This AI development of BGs is finding applications in robotics and computer aided instruction.

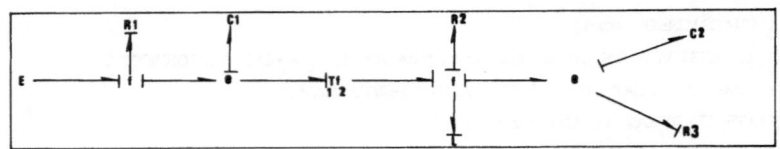

```
STATE_VAR = (QC1, QC2, PL)
STATE_TO_TREAT = (QC1, QC2, PL)
CURRENTLY_ASSIGNED = (ET, QC1, QC2, PL)
STATE_TO_TREAT = (QC2, PL)
VAR_TO_TREAT = (QC1) ---> UNASSIGNED = (FC1)
FC1 = F1 - FTF1
CURRENTLY_ASSIGNED = (ET, QC1, QC2, PL, FC1)
UNASSIGNED = (empty)
UNASSIGNED = (F1, FTF1)
F1 = FR1
FTF1 = FTF2 / n
CURRENTLY_ASSIGNED = (ET, QC1, QC2, PL, FC1, F1, FTF1)
UNASSIGNED = (empty)
UNASSIGNED = (FR1, FTF2)
FR1 = VR1 / R1
FTF2 = FL
CURRENTLY_ASSIGNED = (ET, QC1, QC2, PL, FC1, F1, FTF1, FR1, FTF2)
UNASSIGNED = (empty)
UNASSIGNED = (VR1, FL)
VR1 = ET - V1
FL = PL / L
CURRENTLY_ASSIGNED = (ET, QC1, QC2, PL, FC1, F1, FTF1, FR1, FTF2, VR1, FL)
UNASSIGNED = (empty)
UNASSIGNED = (V1)
V1 = QC1 / C1
CURRENTLY_ASSIGNED = (ET, QC1, QC2, PL, FC1, F1, FTF1, FR1, FTF2, VR1, FL, V1)
UNASSIGNED = (empty)
UNASSIGNED = (empty)
VAR_TO_TREAT = (empty)
CURRENTLY_ASSIGNED = (empty)
STATE_TO_TREAT = (PL)
VAR_TO_TREAT = (QC2) ------> UNASSIGNED = (FC2)
.............................................................
```

Fig. 12: MODEL formulation algorithm. This algorithm and the sequential causality algorithm may be relaxed to deal with causal conflicts and implicit loops.

36

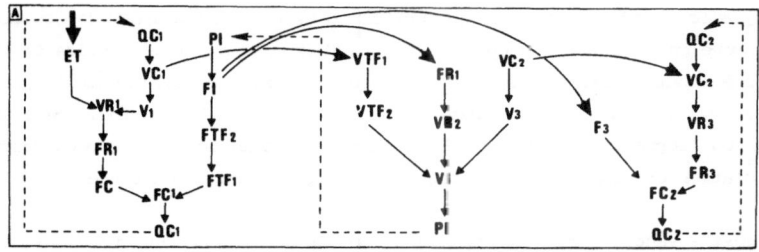

Fig. 13: precedence trees of the MODEL routine

9. CONCLUSIONS

Our brief survey of BGs elementary foundations has only scratched the field : theoretical advanced developments have been overlooked and all significant applications
have not been demonstrated . Modern developments occur roughly in two directions :
  (i) augmentation of descriptive power in new domains or system
  (ii) creation of domain independent general theoretical and practical tools.
As a conclusion, we would only mention some of these developments;

* In multibond graphs [BREE86] , the bonds, instead of transporting a single power
term e * f transmit vectors of efforts and flows . Multiports are defined and n-
dimensional functions and matrix algebra replace the simple scalar functions used
here . 3D-mechanics is the main application of this extended theory .   Its possible
applications in biomechanics are thus obvious .   In addition , BG -applications in
problems usually represented by partial difference equations (e. g. compressible
flow, convection, modal analysis) appear now regularly.

* In network thermodynamics, we have only presented elementary notions. A detailed
treatment of entropy flow may be done including entropy convection, energy theorems
extremum principles and thermodynamic inequalities and couplings, see in [BREE86].
However, BGs have not given many new basic results in the study of non-equilibrium

thermodynamics and they are often considered by the thermodynamics community as some " more mathematical Mickey Mouse " . This does not make justice to what BGs really are about . Being lumped , they just approximate real physical phenomena. Therefore, we should not expect basic " fundamental results ". Their real and , in our opinion, impressive power is in the unified, pragmatic but rigorous description and simulation of complex, multienergetical, highly structured systems.

* The most advanced simulation packages for BGs are the current versions of N-PORT and THTSIM [MEER81] as well as CAMP [GRAN85] which is a BG preprocessor for ACSL Recently , we have also completed a BG preprocessor for our IBM based simulation package SESAME [LEFE86] which is now able to accept and couple graphical modular descriptions of models using IPMs , IBGs , SBGs and BDs as defined in this paper . BDs may be also defined by procedural listings and the system is extensible (with libraries of components and predefined submodels).

TABLE I: correspondance between some physical domains

physical domain	displacement q	effort e = dE/dq	flow f= dq/dt
* translation potential	length or displacement x	force F	velocity
* hydraulic potential	volume V	pressure P	volume flow
* translation kinetic	momentum p	velocity v	force
* hydraulic kinetic	pressure Γ momentum	volume flow f	pressure
* chemical potential	species mole number N	chemical μ	molar flow N'
* thermal	entropy S	temperature T	entropy flow S'
* electric	charge Q	voltage U	current
* magnetic	magnetic flux $\emptyset$	current I	voltage

References

[BREE81] Breedveld, P.C.: Thermodynamic Bond Graphs. Int. J. Modelling and Simulation, 1, 57-61 (1981)

[BREE86] Breedveld, P.C.: A systematic method to derive Bond Graphs models. in: Proceedings of the 2nd european Simulation Conference, 241-248 Eds.: G.C. Van Steenkiste and E.J.H. Kerkhoffs, Antwerpen, SCS (1986)

[GRAN85] Granda, J.J.: Computer generation of physical system differential equations using bond graphs. J. Franklin Institute, 319, 243-255 (1985)

[HOGA87] Hogan, N.: Modularity and Causality in Physical system modeling. Trans. ASME, J. Dyn. Syst; Meas. & Control, 109, 384-391 (1987)

[KARN75] Karnopp, D.C., Rosenberg, R.C.: System dynamics : A unified approach. New York: Wiley (1975)

[KARN81] Karnopp, D.C., Azarbaijani, S.: Pseudo Bond Graphs for generalized compartmental models in Engineering and Physiology. J. Franklin Inst., 312, 95-108 (1981)

[LEFE85] Lefèvre, J., Barreto, J.: A mixed block-diagram and Bond Graph approach to Biochemical systems with mass action and rate law kinetics. J. Franklin Inst., 319, 137-156 (1985)

[LEFE86] Lefèvre, J., Fabri, R., Roucou, D., Drème, T., Filali, S.: An authoring system coupling simulation to computer aided instruction. Proceedings IMACS-IFACS (eds. Borne, P., Tzafestas, S.L.) Lille, 511-514 (1986)

[MEER81] Meerman, J.W.: THTSIM, software for the simulation of continuous dynamic systems on small and very small computer systems. Int. J. Model. and Simul., 1, 52-56 (1981)

[OSTE73] Oster, G.F., Perelson, A.S., Katchalsky, A.: Network Thermodynamic : dynamic analysis of biophysical systems. Quart. rev. Biophys., 6, 1-134, (1973)

[DIXH77] van Dixhoorn, J.J.: Simulation of Bond graphs on minicomputers. Trans. ASME, J. Dyn. Syst; Meas. & Control, 99, 9-14 (1977)

SIMULATION OF TYPICAL PHYSIOLOGICAL SYSTEMS

Thomas G. Coleman and William J. Gay
Department of Physiology and Biophysics
University of Mississippi Medical Center
Jackson, MS, 39216-4505, USA

An example of a mathematical model of a typical physiological system is presented in this chapter. The example is followed by a description of several mathematical tools which we have found to be useful in the study of such systems.

Physiological models and the strategies used to study them seem to be influenced by both the physiological questions being asked and a variety of technical, economic and social considerations. Some of these considerations are outlined at the beginning of this chapter.

The word *simulation* will be used herein to describe the building of mathematical models of mostly continuous systems and the solving of the resulting family of algebraic and differential equations. Digital computers play a prominent role.

RECENT ADVANCES IN DIGITAL COMPUTING

Progress in simulation has always been dependent on the tools which were available to solve the relevant equations. The historical sequence has been: no tools, general solutions of special forms, numerical methods implemented with pencil and paper, electronic analog computers, and now numerical methods implemented on digital computers.

There are many important mathematical models which do not yield general solutions. This means that the advent of the digital computer offered, for the first time, solutions to families of equations which were heretofore unsolvable. The advent of the microcomputer decreased the cost and increased the scale of opportunities in the biomedical sciences.

Biomedical scientists have generally not had easy access to mainframe computers housed in centralized computing centers. Thus, advances in large computer technology have not had a widespread impact on biomedical simulation. The invention and subsequent rise to prominence of the microcomputer, on the other hand, is potentially very important because it means that virtually all scientists have equal access to some form of digital computer services.

An important question in the early years of the microcomputer was whether or not these machines were powerful enough to solve meaningful mathematical models in a timely fashion. The answer for the most part was *no*. But, the answer is now unequivocally *yes*.

Advances in microcomputer technology in the past few years has been remarkable. Forthcoming advances appear to be just as, or even more, remarkable. The full ramifications of these advances to the biomedical sciences is not understood at this time, as least by these authors, but they all seem to benefit the simulation of biomedical phenomena.

To date, hardware advances appear to overshadow software advances. Specifically, a complete, relatively powerful digital computer can now be configured from just a few high-density integrated circuit packages. Calculations requiring decimal (floating point) arithmetic are completed using high-speed, accurate coprocessors. Memory is very inexpensive and large-memory configurations, relative to a few years ago, are now economically attractive. Graphical displays have improved in every conceivable figure of merit. None of these advances have been unreasonably costly.

The numerical solution of families of algebraic and differential equations in a straightforward and timely manner is an essential part of biomedical simulation. Traditional digital computing methodology may have presented the most serious barrier to progress in this area in the form of limited access and excessive costs, formidable hardware and software user interfaces, and limited computing power. While the biomedical community did not directly stimulate any of the recent technological advances, it can benefit from them. These advances appear to elevate biomedical computing in general and biomedical simulation in particular from being an elitist activity to one which is readily available to all biomedical scientists.

SUCCESS MAY CAUSE TROUBLE

Rapid advancement of microcomputer technology might actually be leading us into some serious problems. This ironical view comes from argument that computer software was ill prepared for the recent advances in computer hardware. Traditionally, the normal user of a large-scale, general purpose computer is someone who is specifically employed to be an expert in digital computer use. The normal user of a modern microcomputer is someone who is specifically paid to be an expert in something other than digital computer use. The biomedical scientist is, of course, a member of this latter group.

A second consideration is that many computer programs and computer systems are designed to be used by persons who use the system for only one purpose. But, in modern computing we might expect a scientist or any other computer user to use a computer for dozens of different tasks in a single day. We might even expect that this typical user will use many different computers. In biomedical research, one of the day's tasks might be model building and testing.

User-computer interaction is a likely source of trouble. This interaction is often called the *user interface.* A traditional source of trouble results from the fact that many programs (e.g., word processors) expect text to be input by the user and, accordingly, they reserve a-z, A-Z, 0-9 for textual input. Additional input from the user comes from additional keypresses and combinations of keypresses, since the primary alphabetical characters are reserved. One popular style of microcomputer also offers special non-alphabetical keys called function keys. Most software for this style of computer expects auxiliary input from the user in term of function key presses and combinations of keypresses featuring the ⟨Alt⟩ and ⟨Ctrl⟩ keys. The meaning of such keypresses is usually neither logical nor intuitive. Further, the use of special keys varies from one program to another. To make a bad situation worse, the location of special keys on the keyboard varies from one computer brand to the next.

The best that a user of multiple programs can do in such an environment is to study the documentation closely and to retain as many of the details as possible. A counterpoint comes from considering the "user interface" of the modern automobile. A person can drive any number of different automobiles in close succession without confusion, without the need for additional training and without documentation. This is because the basic user's interface is universally standardized. Can you imagine the excitement that would be created if each manufacturer had a unique and arbitrary placement for the accelerator, brake and clutch pedals? In contrast, to date we have accepted arbitrary definition of keystrokes in modern computing without objection.

It is encouraging that one manufacturer of microcomputers has had the foresight to design and attempt to maintain a standard interface across both system and applications software. Other manufacturers appear to be moving toward a comparable standardization. We maintain that any forthcoming standardization will be of great benefit to biomedical simulation.

We expect that tomorrow's biomedical scientist will be very selective in acquiring new computer-based tools. Thus, software used in biomedical simulation should be expected to have an appropriate user interface as well as adequate more technical, characteristics.

OTHER ISSUES

There is more than one use for simulation in the study of biomedical systems. Sometimes simulation is used for demonstration -- to take fairly well established facts and illustrate the deductive consequences of their interaction. The mathematical model 'HUMAN' is an example (Coleman and Randall, 1983). These models can be thought of as the biomedical equivalent of aviation's flight trainers.

A second use of simulation is related to the conduct of research and the nature of scientific method. A mathematical model may be considered to be a formal statement of a scientific theory that is potentially superior to an equivalent statement using traditional language (Coleman, 1972). Simulations are then thought of as deductive ramifications of the theory and, as such, are an integral part of the application of scientific method to biomedical problems (Coleman, 1975).

Currently, many scientists are feeling strong social pressures to

- use fewer experimental animals in biomedical research,

- make biomedical research more efficient in terms of achieving scientific objectives as directly as possible, and

- make biomedical research more economical in terms of requiring fewer monies to achieve stated scientific objectives.

These considerations introduce the unresolved question of whether or not simulation might be used to greater advantage today in routine biomedical research. As detailed above, many technological barriers have all but disappeared. The argument can be made that simulation is a legitimate tool in the toolbox of scientific method and that doing research without any accompanying simulation is quite possibly inefficient. While this contention will not be (and should not be) universally shared, it is supported by consideration of the complexity often found in biomedical systems, as described below.

There is a growing awareness or admission that biomedical systems are rather complicated as a rule, not as an exception. Offered as evidence is the wealth of peptides which are now known to be neurotransmitters in the central and peripheral nervous systems. Also offered as evidence is the large number of precursors, hormones and metabolites which are active in typical endocrine communication. Thus, in the face of complexity the mathematical statement of theory may gain some advantage over statements using more limited traditional language. The potential utility of simulation might be greater now than in the earlier years when biomedical research in general and physiology in particular were seeking first evidence of the most basic of interrelationships.

The biomedical scientist is now faced with the challenge of building mathematical models which are as accurate and realistic (i.e., isomorphic) as possible. Software developers are faced with the challenge of creating software for popular, inexpensive microcomputers that can solve the resulting equations in a timely fashion without unnecessary complications. It is hoped that the net result should move biomedical science ahead, not backward.

The following section describes a mathematical model which we offer as a typical physiological model. Subsequent sections describe algorithms which we have found useful in the simulation of typical physiological systems. These algorithms are taken from the design of a simulation package, 'DESOLVER' which is currently under development.

A TYPICAL PHYSIOLOGICAL MODEL

The following model will be used as an example of a typical physiological model, acknowledging of course that no physiological model can really be considered to be typical. This model (Guyton and Coleman, 1967) was originally designed to demonstrate several concepts of the long-term control of the mammalian circulation. The model contains a primary relationship involving the kidney, body fluids and basic hemodynamics. There is also a secondary relationship involving blood flow, vascular resistance and arterial pressure.

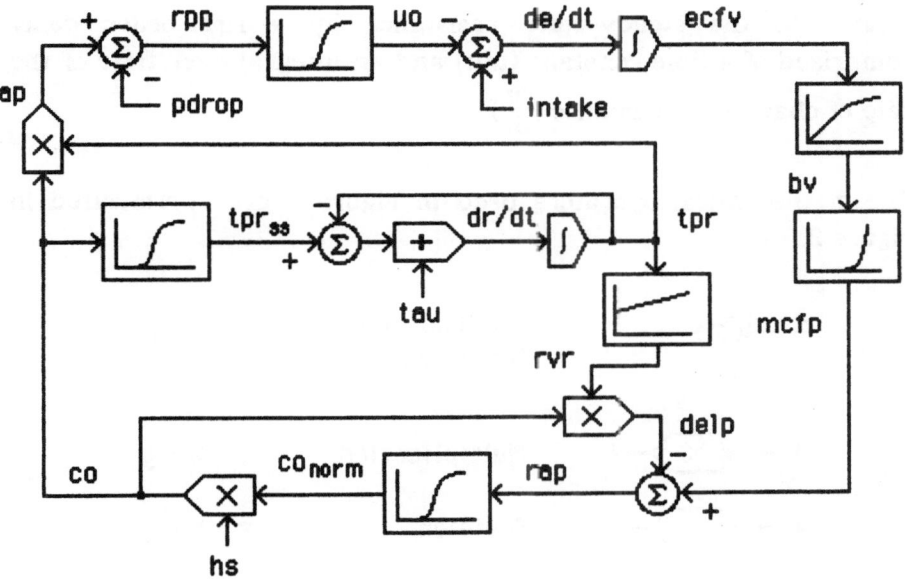

Figure 1

Beginning at the upper left corner of the figure, arterial pressure (ap) is the product of cardiac output (co) and total peripheral resistance (tpr). The renal perfusion pressure (rpp) is equal to the arterial pressure minus the pressure drop in the renal artery (pdrop). Urinary output (uo) is a function of renal perfusion pressure.

The net change in extracellular fluid volume ($\frac{de}{dt}$) is equal to fluid intake (intake) minus urine output. The extracellular fluid volume (ecfv) is the integral over time of the net change in volume. Blood volume (bv) is proportional to extracellular fluid volume and mean circulatory filling pressure (mcfp) is proportional to blood volume.

Right atrial pressure (rap) is equal to the mean circulatory filling pressure minus a pressure gradient (delp) which is the product of cardiac output (co) and the resistance to venous return (rvr).

The steady-state total peripheral resistance (tpr_{ss}) is a function of cardiac output. The immediate total peripheral resistance (tpr) is related to the steady-state resistance by a first-order delay comprised of a time constant (tau) and an integral over time of the rate of change of resistance ($\frac{dr}{dt}$).

The mathematical operators used in Figure 1 are summarized in Figure 2.

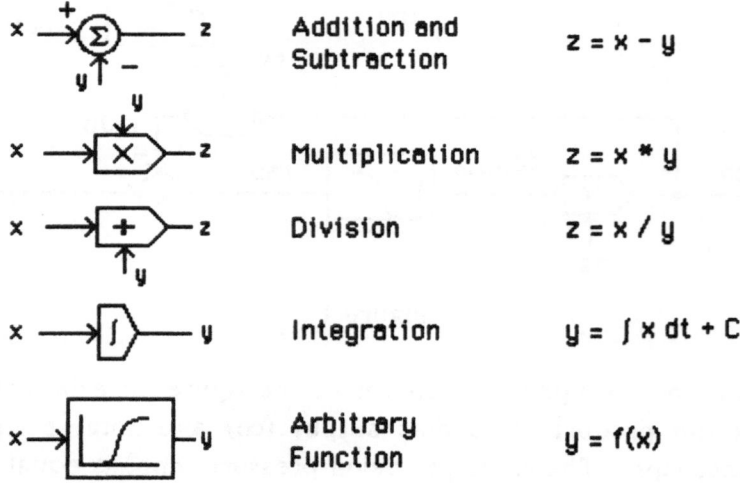

Figure 2

When this model is scaled to the dimensions of a normal human, the physical units and normal values are:

ap	100	mmHg
bv	5000	ml
co	5500	ml/min
co_{norm}	5500	ml/min
delp	7	mmHg
de/dt	0	ml/min
dr/dt	0	mmHg/ml
ecfv	15000	ml
hs	1	x normal
intake	1	ml
mcfp	7	mmHg
pdrop	0	mmHg
rap	0	mmHg
rpp	100	mmHg
rvr	.00127	mmHg/(ml/min)
tau	2000	min
tpr	.0182	mmHg/(ml/min)
tpr_{ss}	.0182	mmHg/(ml/min)
uo	1	ml/min

The model can be used to predict changes in arterial pressure, cardiac output and total peripheral resistance in response to several challenges (Guyton and Coleman, 1967).

In particular, the genesis to renovascular hypertension can be simulated by increasing pdrop from a normal value of 0 to 40 or 50 mmHg as illustrated in Figure 3. This simulation illustrates the recovery of renal perfusion pressure (rpp) with the price being the development of arterial hypertension, as denoted by a gradual increase in arterial pressure (ap). This simulation also illustrates the interaction between body fluids and the control of arterial resistance. The decrease in renal perfusion pressure initially causes fluid retention. The fluid retention causes an increase in cardiac output (co) which tends to elevate arterial pressure. But, the longer-term response shows autoregulatory vasoconstriction, a return of cardiac output toward normal and a sustained increase in total peripheral resistance (tpr). The solution interval is 10000 minutes or approximately one week.

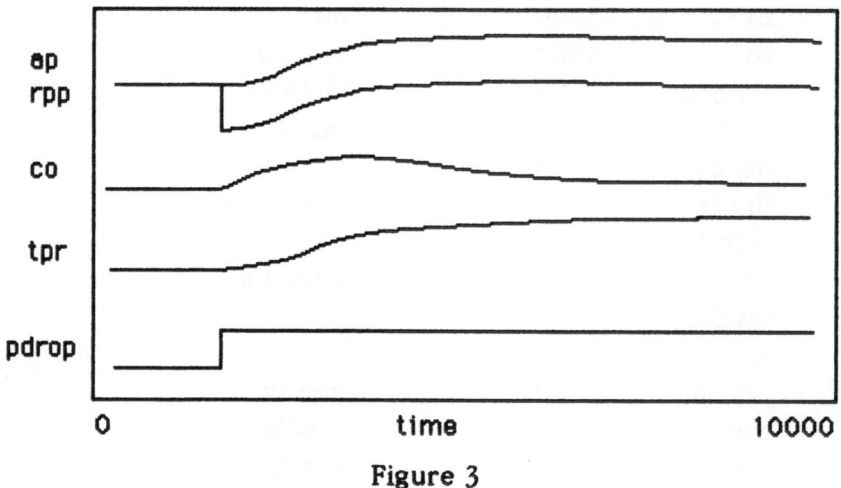

Figure 3

Heart failure can be simulated by decreasing heart strength (hs) from a normal value of 1 to, say, 0.4. The response is illustrated in Figure 4. Acute decreases in cardiac output and arterial pressure are corrected over the longer term by fluid retention and blood volume expansion. The solution interval is again 10000 minutes.

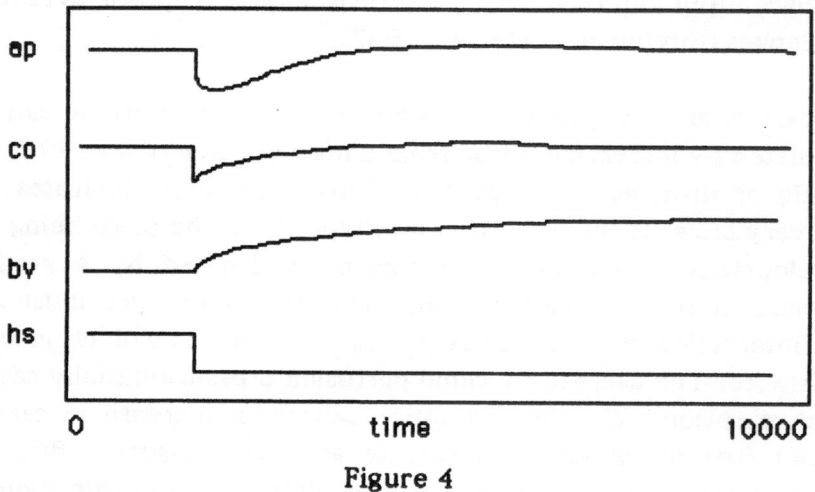

Figure 4

These two responses illustrate the complex, non-linear, time-dependent behavior that is characteristic of physiological models.

This typical model contains several general features which may create problems for the modeller. They are listed below.

- Many algebraic relationships are quite naturally described by curvilinear functions. This model contains six such functions.

- A group of algebraic equations will often contain a non-linear relationship, necessitating iterative solution. In this model, cardiac output (co) is an algebraic function of mean circulatory filling pressure (mcfp), the resistance to venous return (rvr), and the parameter heart strength (hs). The non-linear relationship in this case is cardiac output as a function of right atrial pressure -- the Frank-Starling curve of the heart.

- Either the range of time constants is large or the difference between the shortest time constant and the solution interval is large. The latter is true is this example. The integration algorithm must proceed cautiously, particularly when discontinuities, such as instantaneous changes in heart strength, are possible.

These three characteristics are discussed in following sections along with effective methods of managing the problems which they tend to create.

NON-LINEAR RELATIONSHIPS

Simulation tools often encourage us to build linear models. There are several identifiable reasons:

- linear models are easier to describe,

- linear models are generally much easier to solve, and

- the programming languages which underlie the simulation software are most suitably applied to simple, linear relationships.

In contrast, nature seems to have a tendency to build non-linear systems.

As an example, when programming in a high-level language (e.g., BASIC, Pascal) it is natural to define a linear relationship such as

$$y = k * x$$

where y is the dependent variable, k is a constant and x is the independent variable. If an upper or lower limit is required, the code might be expanded to

```
if (x < k₁) then y = k₂ * x
if (x >= k₁) then y = k₃
```

Descriptions such as these are called piecewise linear. They can be criticized for two reasons. One is that such descriptions are often not a particularly accurate representation of typical (non-linear) biological relationships. The other is that such descriptions can produce a discontinuity in a model's derivatives which will tend, in turn, to disrupt the integration algorithm.

An alternative, which we think is preferable, is to use a smooth, non-linear description. Cubic splines appear to be ideal.

Cubic splines are typically a family of cubic equations arranged in such a way that both the dependent variable and its first derivative are continuous (Press *etal* 1986). Visually, one sees a smooth, curvilinear relationship. Each spline's equation, represented by a domain in the independent variable and a set of coefficients, presents little intuitive value to the modeller. Thus, it seems best to make the graphical representation of cubic splines highly accessible while hiding the coefficient set.

These considerations suggest the following curve definition paradigm.

- Model builder defines the curve in terms of the x,y locations that the curve should pass through and the slope of the curve at each of these locations.

- Computer generates the coefficient set for the cubic splines but keeps it hidden.

- Computer shows model builder a graph of the curve. Model builder judges adequacy. If adequate, go to next step. If not adequate, go back to first step.

- Computer uses cubic splines coefficient set in solving model.

An arbitrary curve ($y - f(x)$) is easily specified as a series of (x, y) pairs with an additional specification of slope ($\frac{dy}{dx}$) at each data pair. The curve is divided into segments running from one data pair to the next. Each segment will be described by a single spline. The entire curve will be described by a family of splines in series.

Suppose we want a segment of a curve to pass through (x_1, y_1) with a slope of s_1 and (x_2, y_2) with a slope of s_2. A polynomial is required and a cubic will give an exact fit. Let

$$y = c_0 + c_1 * x + c_2 * x^2 + c_3 * x^3$$

The slope of y with respect to x is then

$$\frac{dy}{dx} = c_1 + 2 * c_2 * x + 3 * c_3 * x^2$$

When our segment or spline specifications are inserted in these two equations, the result is four simultaneous equations which can be solved for c_0 ... c_3 using regular matrix mathematics.

$$y_1 = 1 * c_0 + x_1 * c_1 + x_1^2 * c_2 + x_1^3 * c_3$$
$$s_1 = 0 * c_0 + 1 * c_1 + 2 * x_1 * c_2 + 3 * x_1^2 * c_3$$
$$y_2 = 1 * c_0 + x_2 * c_1 + x_2^2 * c_2 + x_2^3 * c_3$$
$$s_2 = 0 * c_0 + 1 * c_1 + 2 * x_2 * c_2 + 3 * x_2^2 * c_3$$

Thus, each spline in the family of splines is defined by a domain of applicability (in x) and a four-part coefficient set. During a numerical solution, this data is used to calculate the value of y when given a value of x. Higher-order computer languages which support arrays are very suitable for implementation.

The total number of splines in a family is arbitrary. Many biomedical functions can be adequately described using only a few splines.

Many physiological relationship are sigmoidal or $\int$ shaped. The traditional cubic splines approach can be modified slightly to accommodate this fact. We make, by default, y equal to a linear function of x passing through (x_1, y_1) with slope of s_1 when x is less than x_1 and y equal to a linear function of x passing through (x_n, y_n) with slope s_n when x is greater than x_n, where n is the subscript of the largest x in the data set used to define the splines. These concepts are illustrated below in Figure 5.

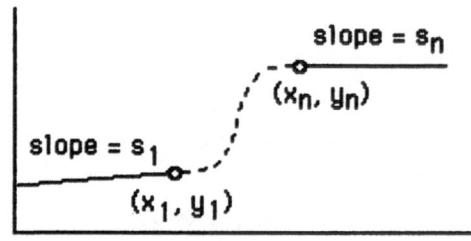

Figure 5

With this default feature, many popular curves can be defined with one or two central splines and a pair of linear tails.

As a concrete example, consider the relationship between right atrial pressure (rap) and normal cardiac output (co_{norm}) used in the model shown in Figure 1 and isolated here in Figure 6.

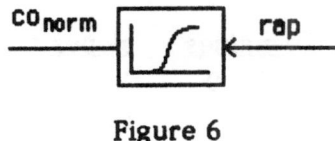

Figure 6

This rendering of the classical Frank-Starling relationship consists of a relatively steep and linear mid-portion with flat right-hand and left-hand tails. The normal operating point is about 1/3 of the way up the curve and in the linear region. For a normal adult human, values of 0 mmHg for right atrial pressure and 5500 ml/min for cardiac output are typical. A simulation which uses only small excursions from this normal operating point will require only a linear representation of the Frank-Starling relationship of the form:

$$co = k_1 * rap + k_2$$

where k_1 is a constant of approximately 1250 (ml/min)/mmHg and k_2 is a constant of about 5500 ml/min.

However, it is the performance of the heart at the extremes which is of interest in a great many cases. First, the heart will not pump properly when right atrial pressure is more negative than than -3 or -4 mmHg since filling is significantly impaired at negative pressures. Thus, we need a flat left-hand tail showing zero flow. Second, increases in right atrial pressure produce increases in cardiac output only when right atrial pressure is relatively low and the ventricles are not distended. At higher pressure levels, pressure increases do not augment cardiac output because cardiac filling and the vigor of cardiac contraction are already close to their maximum values. Thus, we needed a nearly flat right-hand tail showing maximum flow. It is this right-hand tail which determines maximum tolerance to exercise in normal subjects and tolerance to any physical activity at all in heart-failure patients.

A typical cardiac function curve might be defined using these data points.

x-values	y-values	slopes
-4	0	0
0	5500	1250
12	12500	10

The data points and resulting splines and tails are illustrated below in Figure 7.

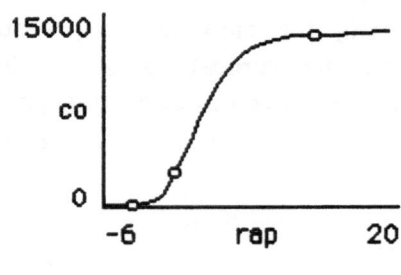

Figure 7

ALGEBRAIC RELATIONSHIPS

A typical physiological model contains differential equations in need of integration and algebraic expressions which define the derivatives. These algebraic relationships sometimes assume the implicit form of $x = f(x)$.

If an implicit algebraic relationship is simple and linear, an explicit solution can generally be found and embedded in the model. If a relationship is moderately simple and moderately linear, matrix methods can be used to calculate the numerical solution to the

simultaneous equations. Otherwise, an iterative solution is
indicated. Iterative solution is required, for instance, when cubic
splines are embedded in an implicit algebraic relationship.

A traditional example of this third case is found in the calculation of
cardiac output as a function of both cardiac and peripheral factors,
as illustrated in the model in Figure 1 and isolated here in Figure 8.

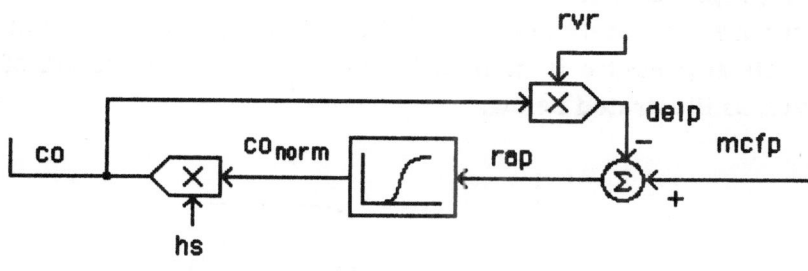

Figure 8

This problem has traditionally been solved graphically by finding
the intersection of the cardiac function curve with the venous
return curve (Guyton *etal* 1973) as illustrated below in Figure 9.

The cardiac function curve represents the pumping capability of the
heart in terms of the outflow of blood (ml/min) as a function of the
filling pressure or right atrial pressure (mmHg). This capability,
often called the Frank-Starling curve, is a combination of the basic
cardiac function modulated by the heart strength.

The venous return curve represents the state of the peripheral
circulation in terms of inflow of blood to heart (ml/min) as a
function of the the filling pressure or right atrial pressure (mmHg).
The shape of the venous return curve is determined by the mean
circulatory filling pressure (mmHg) and the resistance to venous
return (mmHg/(ml/min)). Specifically, the mean circulatory filling
pressure determines the intersection of the venous return curve
with the horizontal axis and the resistance to venous return
determines the slope.

Right atrial pressure is a variable in both the cardiac and peripheral systems. At equilibrium, the cardiac outflow (cardiac output) must equal cardiac input (venous return). Given values for all of the relevant parameters, finding a solution consists of calculating the value of right atrial pressure which will yield equal values of cardiac output and venous return.

For a graphical solution, the cardiac function and venous return curves are accurately drawn and the prevailing cardiac output and right atrial pressure is defined by the point of intersection of the curves, as illustrated below.

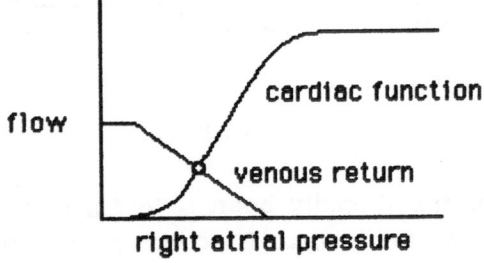

Figure 9

Two options for computerized solution will be considered. One is to linearize both the cardiac function and venous return curves. An explicit solution follows directly. Linearization can be expected to impair the model's performance, however, particularly in instances of high venous return or decreased cardiac function. Thus, linearization is often not acceptable. A second option is to retain a curvilinear cardiac function curve, linearize the venous return curve and use iterative numerical methods to find the solution. This second approach is a bit more versatile than the first approach.

The relevant variables and parameters are

co	cardiac output
vr	venous return
rap	right atrial pressure
mcfp	mean circulatory filling pressure
rvr	resistance to venous return
k_1, k_2	cardiac function parameters

A linear representation of cardiac function was developed in the previous section.

$$co = k_1 * rap + k_2$$

Venous return is the ratio of the pressure gradient promoting flow and the resistance opposing it.

$$vr = \frac{mcfp - rap}{rvr}$$

Since co and vr are identical, these two equation can readily be solved for rap and, subsequently, co.

If we maintain the cardiac function curve as a curvilinear function, then

$$co = curve\ (rap)$$

where curve represents an arbitrary function possibly represented by a set of cubic splines, and

$$vr = \frac{mcfp - rap}{rvr}$$

Again noting that co and vr are identical, we combine these two equations to obtain

$$co = curve\ (mcfp - co * rvr)$$

This resulting equation is now in need of solution. Since **curve** is an arbitrary function, no general solution is possible. Instead, an iterative technique can be used to calculate a numerical solution.

Before we proceed, notice in the example above that **rvr** and the slope of **curve** are expected to be positive. If so, then the partial derivative of the right-hand side of this equation with respect to **co** is negative. This is good. When bisection is used, the first two estimates of the root must bracket the root. A negative derivative insures bracketing. More on this below.

The general iterative technique for solving implicit algebraic equations is as follows. Combine the equations to obtain a single equation

$$g(x) = 0$$

where $g(x)$ has the form

$$g(x) = x - f(x)$$

and then solve for x. If the equation has multiple or complex roots, then trouble is at hand. Fortunately, such equations are not common in physiological research. If the equation has discontinuities, then be advised that many iterative techniques may have trouble finding the root.

These considerations may make it sound like algebraic relationships with non-linear components are an invitation to disaster, but the opposite is actually true. In our experience, the majority of physiological examples fall into a single, easily solved category. The key is this: if the derivative of $f(x)$ with respect to x is negative for all x, then the following algorithm holds.

- Given $g(x) = x - f(x) = 0$. Pick an arbitrary value for x, called x_1. Let $x_2 = f(x_1)$.

- Values x_1 and x_2 now bracket the solution of $x - f(x) = 0$ and bisection can be used (Press *etal* 1986). Bisection always converges. The anxiety is gone.

Bisection works as follows. Consider the implicit equation to be

$$x - f(x) = \Delta$$

Then x is declared to be a root when the absolute value of Δ is sufficiently small.

Beginning with values for x_1 and x_2

- Use x_1 to calculate Δ_1. Use x_2 to calculate Δ_2. (We might test Δ_1 and Δ_2 here. If either is sufficiently small, we declare the appropriate x to be the root and stop.)

- Calculate $x_3 = \dfrac{x_1 + x_2}{2}$. Calculate Δ_3. If Δ_3 is sufficiently small then declare x_3 to be the root and stop.

- If Δ_1 and Δ_3 have the same sign, replace x_1 and Δ_1 with x_3 and Δ_3. Go to step 2.

- Δ_1 and Δ_2 always have opposite signs. If the conditions of step 3 are not met, then replace x_2 and Δ_2 with x_3 and Δ_3. Go to step 2.

Continue until convergence is detected at step 2. Other methods may be faster than bisection, but bisection is suitable for many physiological models. It's reliable, predicable and easily implemented.

INTEGRAL RELATIONSHIPS

Integral relationships are a fundamental part of most physiological models, including the typical model presented in Figure 1. The two integral relationships in this model are isolated here as Figures 10 and 11.

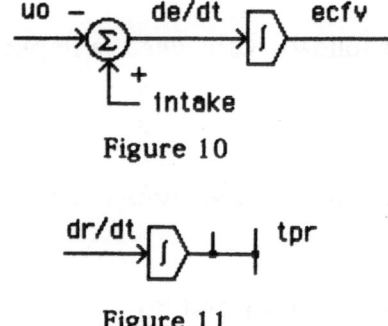

Figure 10

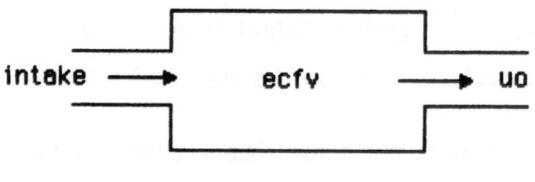

Figure 11

In the case of extracellular fluid volume, and many other physiological relationships, it is an aspect of mass balance which is being modelled. A schematic view is shown below.

intake ⟶ ecfv ⟶ uo

Figure 12

The corresponding equations are:

$$\frac{de}{dt} = intake - uo$$

$$ecfv = \int_{0}^{t} \frac{de}{dt} * dt$$

The independent variable is t and its differential is dt.

The usual condition is that the derivative is not explicitly defined and, therefore, numerical integration is used iteratively to calculate the integral. But, numerical integration is only an approximation of true integration. This means that numerical integration is inherently in error -- sometimes to an unacceptable degree. The ever present threat that numerical integration error will invalidate a solution has stimulated a considerable effort in many scientific and other areas to develop suitable integration algorithms.

The easiest approach to numerical integration is to use a first-order Euler algorithm. Using the extracellular fluid volume compartment described above, first-order Euler integration requires a difference equation which can be expressed casually as

$$\text{new\_ecfv} = \text{old\_ecfv} + \frac{de}{dt} * dt$$

The charm of this algorithm is its simplicity. The shortcoming is that there is no estimate of the prevailing error. The algorithm produces both acceptable and highly erroneous results with equal facility.

Toward the other extreme, there are rather complicated integration algorithms which we will label here as higher-order (see Press *etal* 1986 for examples). In the example above, $\frac{de}{dt}$ is the first derivative of ecfv and it is the only derivative used in the first-order Euler algorithm. The higher-order integration algorithms, in contrast, use estimates for higher derivatives, error detection, and a strategy for error control using trial or repeated calculations.

Higher-order schemes are a mixed blessing. They can be very efficient; undoubtedly, this was particularly important when numerical solutions were implemented with pencil and paper before the advent of the digital computer. It is still an important consideration. One negative consideration is that higher-order algorithms do not respond well to discontinuities in the derivatives. This is relevant to physiological modelling where it is often useful to produce an abrupt change in one or more of the parameters and then to observe the transient response of the model. And, every solution is likely to have a discontinuity in derivatives occurring at the beginning of the solution. Thus, the implementation of higher-order integration schemes requires attention to detail which is justified in many special cases but may not be necessary for many typical physiological models.

Lying between the extremes presented above is an algorithm which we have found to be easy to implement and dependable in use. It is a variable step-size, first-order algorithm. A brief derivation follows.

In numerical integration, the independent variable is advanced in small steps called the integration interval, which is often referred to as dt. Given the value of a derivative and its integral at the beginning of an integration interval, the challenge is to find the value of the integral at the end of the interval.

Taylor's series adds perspective. If the independent variable is t_0 at the beginning of the interval and $x(t_0)$ is the value of the integral at t_0 then

$$x(t_0+dt) = x(t_0) + \frac{\frac{dx(t_0)}{dt} * dt}{1!} + \frac{\frac{d^2x(t_0)}{dt^2} * dt^2}{2!} + \dots$$

$$\dots + \frac{\frac{d^n x(t_0)}{dt^n} * dt^n}{n!} + \dots$$

With the proper amount of information we can calculate $x(t_0+dt)$ exactly. The proper amount of information is all of the higher-order derivatives of x, and it is safe to conclude that this information is not available. As previously stated, the challenge is to estimate $x(t_0+dt)$ when given $x(t_0)$, $\frac{dx(t_0)}{dt}$ and dt.

Notice what happens to Taylor's series when we declare all derivatives except the first to be unimportant. The unimportant

terms are discarded and Taylor's series reduces to the first-order
Euler algorithm previously presented. The discarded terms are, in
fact, the integration error. Thus, we might write Taylor's series as

$$x(t_0 + dt) = x(t_0) + \frac{dx(t_0)}{dt} * dt + error$$

If we now assume that the first of the truncated terms is the
largest contributor to the error, we can estimate the error as

$$error \approx \frac{\frac{d^2x(t_0)}{dt^2} * dt^2}{2}$$

This is a rough estimate of the error and not an upper bound.
Notice, however, that the error can be made arbitrarily small by
adjusting dt.

While the second derivative of x is generally not explicitly
available, it can be estimated near t_0 and included in the formula
for error.

$$error \approx \frac{\frac{\frac{dx(t_0)}{dt} - \frac{dx(t_0 + dt)}{dt}}{dt} * dt^2}{2}$$

Finally, we must consider both the sign and the relative magnitude
of the error. The error estimate above shows that both positive and
negative integration errors can occur. Since these two error types
are equally undesirable, the absolute value of the error should be
used in normal calculation. Further, the seriousness of any
integration error must be weighed against the maximum tolerable
error (max_err). The maximum tolerable error is specified by the
model builder. These two considerations suggest that a relative
error be used.

$$\text{relative\_error} = \left| \frac{\text{error}}{\text{max\_err}} \right|$$

Thus, the integration algorithm.

- Begin with $x(t_0)$, $\dfrac{dx(t_0)}{dt}$ and dt.

- Calculate $x(t_0{+}dt)$ using Taylor's series through only the first derivative (i.e., integrate using first-order Euler).

- Calculate $\dfrac{dx(t_0{+}dt)}{dt}$ from x $(t_0{+}dt)$. Use the error equation above to estimate the error. If error is acceptable, then we are finished with this interval. If error is too large, then decrease dt and return to step 1.

This algorithm reduces the chances that large integration errors will occur. However, it increases the chances that an unacceptably small dt will be created. A small dt protects against integration error but also extends the total calculation time.

The challenge is to find a dt which is large but not too large. At any point in the solution, we can estimate the value of dt (**max_dt**) that will exactly yield the maximum tolerable integration error (**max_err**). This dt, or some fraction of it, can then be used as an efficient dt.

Focus only on the integral that is producing the largest relative integration error. Define **k** as the absolute value of the current second derivative of this integral divided by 2. The estimate of integration error at **max_dt** is

$$\text{max\_err} = k * \text{max\_dt}^2$$

Then, solving for **max_dt**

$$max\_dt = \sqrt{\frac{max\_err}{k}}$$

Since the absolute value of the current second derivative divided by 2 (k), the current dt (cur_dt) and the current error (cur_err) are all related by

$$cur\_err = k * cur\_dt^2$$

we can replace k in the estimate of max_dt to obtain

$$max\_dt = cur\_dt * \sqrt{\frac{max\_err}{cur\_err}}$$

We use this final equation for max_dt to get a new estimate of dt whenever error is either too large or too small. The accuracy of this calculation is only as good as the estimate of integration error, which depends, in turn, on the suitability of using the second-order term in Taylor's series as an estimate of integration error.

Once the maximum allowable error is specified for each integral, the integration algorithm described above (and others like it) is maintenance free. This prospect can be very reassuring to anyone whose primary skills are outside of modelling.

SUMMARY

The simulation of physiological systems has the potential to become an important part of physiological research. Current methodology employs digital computers and numerical methods.

The advent and continuing development of inexpensive microcomputers has solved an important technological problem. A new problem of software complexity has appeared. Software, including simulation software, that is difficult to use will not be accepted and used by the busy scientist. When these problems are overcome, we can expect simulation to become a routine part of biomedical scientific method, particularly when complex processes are involved.

The long-term control of the circulation was used to illustrate some features of a typical physiological model. Each of the highlighted features required an appropriate numerical algorithm for use in calculating solutions.

One feature was non-linear relationships. Non-linear relationships are common in nature and it was recommended that they should be preserved in the model building and model solving process. Cubic splines are one of many techniques that can be used. They are particularly suited to digital computation.

A second feature was implicit algebraic relationships. When an algebraic relationship contains a non-linearity, an iterative method is required. The method of bisection was recommended.

A third, and very common, feature was integral relationships. Some care must be taken to obtain stable, accurate solutions to families of differential equations. While many suitable numerical methods are available, a first-order, variable-step-size algorithm was described and recommended because of its simplicity.

ACKNOWLEDGEMENT

This work was supported in part by United States National Institutes of Health grant HL 11678.

REFERENCES

Coleman, T.G. From Aristotle to modern computers: The role of theories in biological research. Physiologist. 18:509-518, 1975.

Coleman, T.G. Simulation is helping biomedical research. Simulation. 19: (Simulation Today) 29-32, 1972.

Coleman, T.G. and J.E. Randall. HUMAN - A comprehensive physiological model. Physiologist. 26(1): 15-21, 1983.

Guyton, A.C. and T.G. Coleman. Long-term regulation of the circulation: Interrelationships with body fluid volumes. In: Physical Bases of Circulatory Transport: Regulation and Exchange. E.B. Reeve and A.C. Guyton (Eds.) W.B. Saunders, Philadelphia, pp. 179-201, 1967.

Guyton, A.C., C.E. Jones and T.G. Coleman. Circulatory Physiology: Cardiac Output and its Regulation. W.B. Saunders, Philadelphia, 1973.

Press, W.H., B.P. Flannery, S.A. Teukolsky and W.T. Vetterling. Numerical Recipes. Cambridge University Press, Cambridge, 1986.

Parameter estimation: an advanced simulation tool in biomedicine

Dietmar P. F. Möller

Physiologisches Institut, Universität Mainz
Saarstr. 21, D-6500 Mainz, W.-Germany

Introduction

Parameter estimation as an advanced tool in biomedicine have become extremely important in the study of biological systems. With increasing physiological knowledge and more efficient computers, modelling, simulation, and estimation have grown during the past 20 years to one of the most powerful tools of biomedical system analysis. Moreover modelling, simulation and estimation is of growing importance for determination of non-measurable parameters and state variables of biomedical systems as well as for case studies research of biomedical processes, which have been developed by physicians and system engineers and have also been widely accepted by biomedical scientists. These models already developed have been applied for an extensive analysis of dynamic transient behaviour under different conditions by systems simulation. In this way the most important physiologically relevant system parameters have been very often estimated from measurements obtained from experiments. However, modelling and simulation of biomedical systems is in general the inverse of the simulation process in engineering science and systems theory. For the biomedical systems have already existed and have been optimized by evolution.

Under normal circumstances, the biomedical scientist is not solely interested in the mathematical model of the biomedical system under normal -in medical terms healthy- or pathological conditions. He might prefer the developed mathematical model adequately describe the pathological behaviour in the case of diseases. Hence a disease can roughly be stated as a system behaviour outside normality.

71

In contrast, the main goals in engineering application are systems synthesis and optimization, and the systems engineer is primarely interested in the mathematical model of a process under normal operating conditions. His aim is to use the model in order to control the process optimally or to keep it at least in a relatively close vicinity of conditions that award the danger of a possible drifting of the process to margins of safe operating conditions [3].

The simulation process of a biomedical system is an iterative one, consisting of model building and computer assisted simulation by changing the real structure of the model and its parameters with the aid of parameter estimation in an effort to match the real biological system well. In fact, the derived model has served its purpose when an optimal match is obtained between the simulation results and the data obtained from the real system itself.

In general the model building process of a biomedical system entails the utilisation of two types of information:
- a priori physiological knowledge of the system being modelled (a biomedical system must me observable).
- experimental data consisting of measurements of the system itself.

With respect to the spectrum of available models, a variety of levels of conceptual and mathematical representation is evident, which depends from the purpose for which the model was intended and the extent to the a priori physiological knowledge available.
However, not only from a more general point of view, two major facts are of importance when modelling biomedical systems:

A model always is a simplification of reality, but should never be so simple, that its answers are not true.

A model has to be simple to allow easy handling, working and understanding with it.

These are the two relevant boundary conditions for model building. Because model building is a compromise between model goodness (i.e. the exactess of the results obtained from the models) and the expenditure of modelling (i.e. the costs for developing the model, its implementation on the computer, its simulation and parameter estimation), which is shown in Figure 1.

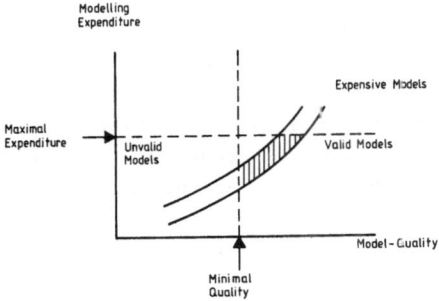

Figure 1: Dependance of the modelling expenditure (costs) versus the
degree of accuracy (model quality).

From Figure 1 one can conclude, that there is no reason to develop
expensive models, because increment of goodness is less than the
increase of costs. This point is of importance since a mathematical
model is a very compact way to describe a biomedical system. Because
a complex model describes not only the relationships between the
system inputs and outputs, but gives also a detailed insight into
the system's structure and into some system internal relationships.
This is due to the fact that the main relationships between the
physical variables of the biomedical system to be modelled are
mapped into appropriate mathematical expressions. For instance, the
relation between input-and-output variables of a system will be
described (depending on its complexity) by an ordinary second order
differential equation or by a set of first order differential
equations, which represent the mathematical model of a biomedical
system.

In principle, there exists two different approaches in obtaining the
system model, e.g. a pure theoretical one based on the derivation of
the essential physical (including the chemical or biochemical)
relationships within the system and a pure empirical one based on
experiments on the system itself. Practical approaches use a
combination of both, which might be the most advantageous.
Concerning the theoretical derivation of model equations for a
physiological system, as for example the balance equations of the
systems must be taken into account, the law of conservation of mass,

energy of momentum, as well as some further phenomenological laws typical of the system under test, and the existing boundary conditions.

It is important to note, that in biomedicine in general, model building is an essential approach for estimation of clinically important but under normal circumstances not directly measurable system parameters and internal system states, or even an essential excitation of the system as required for efficient parameter estimation, is (from the medical point of view) not allowed under clinical conditions.

Model building of the renovascular system

Model building of the renovascular system is generally based on the a priori knowledge about the system itself and on experimental data available from patients or animal experiments.
The spectrum of the published literature in this area includes various models, a mathematical representation of which depends in a great degree of the purpose for which the models have been built.
This paper describes an extended model of the renovascular system which is based on a system model of Guyton and Coleman [2] to study the renal function and especially to explain parameter estimation of non-linear biomedical systems.

Figure 2 shows a simplified block diagram, depicting the relevant inter-relationships between venous return (cardiac output) (V̇R, HZV) total peripheral resistance (RA), arterial blood pressure (PAS), extracellular fluid volume (VECF) and blood volume (VB).

Volume control (plasma volume, extracellular fluid volume) can only work in a sufficient manner if osmotic pressure is kept constant within narrow limits which is almost equivalent to control of sodium balance. The model, shown in Figure 2, acts as follows. The urinary output (UO) increases in case that arterial blood pressure (PAS) increases. This causes a decrease in the body fluid volume (VB) and the mean systemic filling pressure (PMS), also the difference between the right atrial pressure (PRA) and the mean systemic filling pressure (PMS). With regard to the resistance of the venous return (RVR) the venous return (V̇R) decreases and hence the cardiac output (CO, HZV). Taking into account a constant peripheral

resistance (RA) the decrease of arterial blood pressure (PAS) is
followed by an elevated renal fluid retention. In this way the loss
of blood volume is compensated by an increase in plasma volume (VP),
and in an elevated extracellular fluid volume (VECF). This feedback
mechanism consolidates the blood volume (VB), the right atrial
pressure (PAS), the venous return (V̇R), the cardiac output (CO,
HZV), the arterial blood pressure (PAS) and the urinary output (UO).

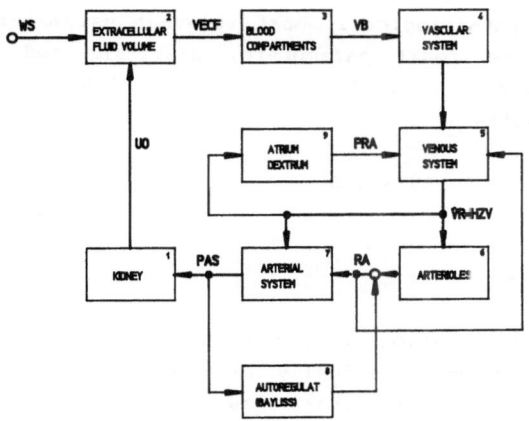

Figure 2: Morphologically oriented structure of the pressure
regulation mechanisms for buffering arterial pressure
when other factors like urine output (UO) or the water
and sodium uptake (WS) change, based on the Guyton-
Coleman model as a special case.

The block diagram form of the model, shown in Figure 2, can be
described in the state-space form by non-linear vector differential
equations as follows:

$$\dot{X}1 = U-F_1 \ (HZV \ \cdot \ X2)$$

$$\dot{X}2 = X3 + X4 = KRA \ (HZV - KHZVF) + RAB$$

HZV represents the cardiac output and KHZVF the cardiac output in
the case of fixed hypertension, RAB represents the myogenic
autoregulation effect, U represents the isotonic sodium and water
net intake, F_1 describes the nonlinearity of the renal function.
X1, X2, X3 and X4 are the state vectors, the components of which are

the extracellular fluid volume (X1), the total peripheral resistance (X2), the metabolic influence on the peripheral resistance (X3) and the myogenic influence on the peripheral resistance (X4).

The non-linear system elements, block 3, 4, 7, and 9, can be expressed by quadratic approximation polynomina.

The model has been used to simulate normal system behaviour as well as pathological behaviours e.g. hypertension cases like Goldblatt-hypertension or reduced renal mass.

The model is also suitable to be used as reference model in a scheme for estimation of physiologically most relevant parameters, in that case these are the renal parameters, a12, a11 and a10, which describe the non-linear renal function curve, which is expressed by the following equation:

$$UO = a12 \cdot PAS^2 + a11 \cdot PAS + a10.$$

Parameter estimation

The main difficulty in parameter estimation of biomedical systems is the large number of physiologically relevant parameters and their interrelation. This requires that some of the system parameters should a priori be known or directly be measured in vivo a requirement that is a rather difficult task.

Another essential problem in parameter estimation of biomedical system is the problem of identifiability, which means the possibility of system parameter estimation based on experiments.

Here, the systems frequently show the internal or structural non-identifiability, so that the identifiability of the given system should be proved usually in the case of linear models by proving the system controllability and system observability, as reported in [3].

If algebraic methods for testing identifiability are not applicable, a graphical test is a proper proof. The philosophy behind this test is that it is important to check whether the parameter estimation based on a large number of initial values, taken from an expected neighbourhood of the system parameter vector will be an estimate of the minimized parameter vector with a sufficient accuracy, as reported in [7], shown in Figure 3.

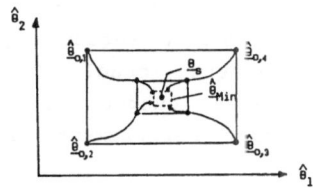

Figure 3: Graphical representation of testing identifiability by
identification [7].

A further essential problem in parameter estimation of biomedical
systems is the problem of parameter estimation accuracy.
Identifiability as discussed before does not imply good accuracy of
the parameter estimates in the presence of measurement noise. If the
variance of the parameter estimates is too big, conclusions from
these estimates will be unreliable. How to get an idea of estimation
accuracy?
As reported in [4], as a measure of the parameter estimation error,
the deviation of the minimum error functional for noise- free data
was chosen, which for low-noise corrupted data gives

$$C_{0V}(\Delta\underline{\theta}_{min}) = \{[\underline{S}\frac{y}{\underline{\theta}}(\underline{\theta}_s)]^T \underline{S}\frac{y}{\underline{\theta}}(\underline{\theta}_s)\}^{-1} \delta_v^2$$

where $\underline{S}$ is a sensitivity matrix of the model output $\underline{y}$ with respect
to $\underline{\theta}$ and δ_v^2 is the variance of the measurement noise. From
this equation it follows that the variances of the estimation errors
can become significant when $\underline{S}$ is bad conditioned. For instance for
relatively small value of δ_v^2 the rough evaluation of the above
equation by calculation of the diagonal elements of the above
equation gives the approximate values of the expected variances of
estimation errors for the respective system parameters. Since the
calculation results are valid only for a small neighbourhood of the
nominal parameter values $\underline{\theta}s$, they have to be tested over an area of
expected true parameter values.

Identification task

In this paper we will assume that the structure of the tested system model describes the system under investigation sufficiently well for the n-dimensional parameter vector θs.

We will speak of the true parameter vector θs and correspondingly of the true model if the output sequence $\{Y_k (\theta)\}$ will coincide with the output of the true model $\{Y_k\}$, if the difference between the measured system output $\{Y_{meas,k}\}$ and the model output $\{V_k\}$ fits the output measurement error Y_k in that way, that

$$\left\{ V_k (\theta) \right\} = Y_{meas,k} - Y_k (\theta) = \left\{ V_k \right\}$$

as shown in Fig. 4.

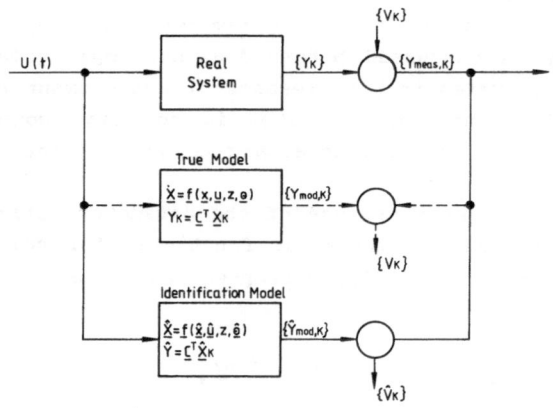

Figure 4: Relationships between the real system, the true model, and identification model [5]

The basic principle of parameter estimation is the adjustment of the parameter vector θ of an identification model, having the same structure as the true model, in such a way that its output $\{yk(\theta)\}$ will coincide with the output $\{yk(\theta s)\}$ of the true model [3,8]. If $\{yk(\theta s)\}$ is not known it is, at most possible to compare $\{yk(\theta)\}$ with $\{y_{meas,k}\}$. If the difference

$$vk(\theta) = y_{meas,k} - yk(\theta)$$

is interpreted as the estimate of vk, the task will be to adjust $\underline{\theta}$ in such a way as to impress on the sequence $\{vk(\underline{\theta})^2\}$ some known statistical properties of $\{vk\}$ [6], e.g. its mean and its variance. Assuming $\{vk\}$ would be white and stationary its mean zero and variance δ_v^2 this task can be done by minimizing the well known output error least-squares functional

$$J(\underline{\theta}) = \sum_{k=1}^{N} v^2 (\underline{\theta}) = \sum_{k=1}^{N} (Y_{meas,k} - Y(\underline{\theta}))^2$$

the minimizing argument of which is a consistant estimate of the true parameter vector $\underline{\theta}s$.

Identification results

Applying the arrangement, shown in Figure 4 which was implemented very first on a PDP 11/45 computer and later on an IBM-PC AT version, the parameters a12 and a11 have been identified by the use of the means of time behaviour of the mean arterial blood pressure under an system excitation by an isotonic water and salt load.

Table 1

Parameter	Estimated Values	Start Values	Nominal Values
a_{12}	3,54	3,6	3,749
a_{11}	5,8	5,8	5,999
a_{12}	3,764	3,39	3,749
a_{11}	6,001	5,66	5,999
a_{12}	3,749	3,8	3,749
a_{11}	6,002	6,1	5,999

Table 1 shows the estimated values of the above parameters a12 and all. It is a wise, that identification of a12 and all as the use of the mean arterial pressure delivers values, which are of great physiological relevance. The presentation of the error functional as a function of a12 or all for n=12 equally spaced measurements are the model step responds, as shown in Figure 5.

The parameters all and a12 in the first row of table 1 show that the minimum of the error function will not be reached. Compared with a second parameter set of all and a12 we found, as a result of Figure 5, an over-estimation of .4 % instead of an under-estimation of 5.7% in the first case. The third row of parameters matches the second one extremely well. In case 2 of the parameter set of all and a12, the error functional is of a form of a deep crater over the two dimensional all and a12 parameter space with an unique minimum for the true parameter values. If however the first set of parameters is assumed, $J_N(\theta)$ is valley-like with a line of largest depth showing a small ascent from an unique deepest point of the true parameter values.

Therefore θs is, again, identifiable in principle but the ascent of the bottom of the value would be zero at all the parameters would no more be identifiable. Practically, it is very difficult to find the search direction along the valley, for instance by the Rosenbrock method, used.

On the other side, gradient methods as the Davidon-Fletcher-Powell-method appear to be not applicable at all because of the non-differentiability of $J_N(\theta)$ along the bottom of the valley. If, during the minimization, a point of this bottom of the bottom of the valley is reached the search hangs up in general.

By the graphical representation of the error functional of the identification results obtained in such cases it can be recognized as point of the bottom edge of the valley [3].

This is a general problem in applying methods for biomedical system analysis, which are well and proper to use for technical system analysis. Especially the minimization of complex biological systems yields problems, which need the interactive handling of the parameter estimation scheme.

How to solve this problem? From a general viewpoint, the development and verification of new software tools for parameter estimation, e.g. specific algorithms must be proceed, like for instance PARFIT

[1], and on the other hand developments in the field of artificial intelligence must be done like the toothing of algorithms into a declarative knowledge base as one integrated artificial intelligence environment, instead of statistical proving, which may be proper to use and an powerful aid, to solve the parameter estimation problem with respect to non-linear biomedical systems.

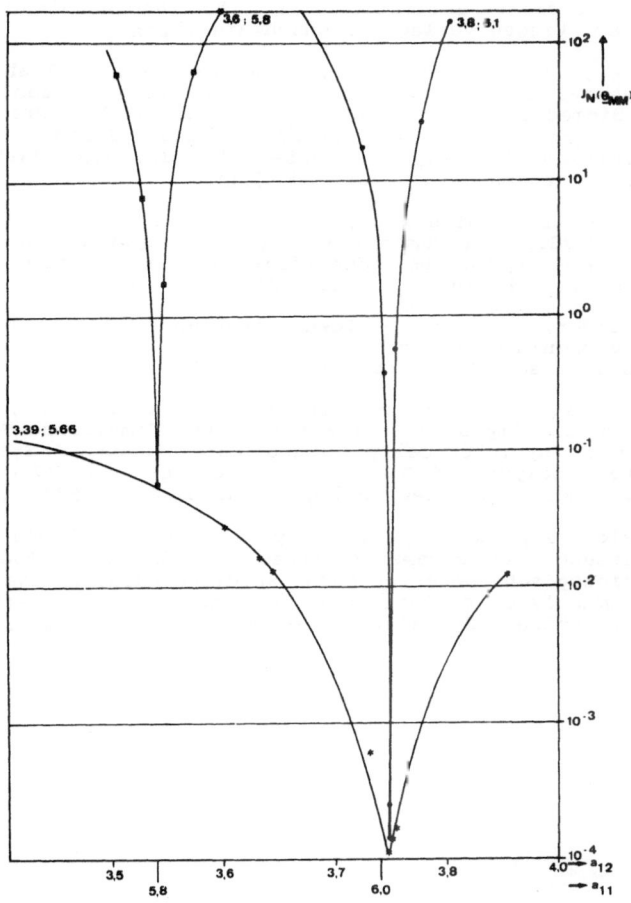

Figure 5: Two-dimensional error-functional $J_N(\theta_{MM})$ which spans a_{12} and a_{11} for n = 12 equally spaced measurements

81

References

[1]   Bock, H.G., Schlöder, J.P.: Fit, Fitter, The Fittest:
        Methods for Modeling and Identification, In: System
        Analysis of Biological Processes, pp. 63-72, Ed.: D.P.
        Möller, Vieweg Verlag, Braunschweig, 1987

[2]   Guyton, A.C.: Arterial Pressure and Hypertension, W.B.
        Saunders Comp., Philadelphia, 1980

[3]   Möller, D., Popovic, D., Thiele, G.: Modeling, Simulation,
        and Parameter-Estimation of the Human Cardiovascular
        System, Vieweg Verlag, Braunschweig, 1983

[4]   Möller, D., Popovic, D., Thiele, G.: Reliability of
        Parameter Estimation Methods applied to the Identification
        of Biomedical Multicompartment Systems, In: Proceed. 7th
        IFAC Symposium Identification and System Parameter
        Estimation, Vol. 2, pp. 1385-1390, Ed.: H.A. Barker, P.C.
        Young, Pergamon Press, Oxford, 1985

[5]   Popovic, D., Möller, D., Thiele, G., Pohl, V., Tanha, A.:
        Case Studies in Human Cardiovascular System, In: Proceed.
        8th IFAC Symposium Identification and System Parameter
        Estimation, Peking, 27.-31.8. 1988

[6]   Schneider, G.: Identifikation bei geringer
        Störgrößeninformation
        Regelungstechnik 27, pp. 110-117, 1979

[7]   Tanha, A., Maftoon, H., Thiele, G., Möller, D., Popovic,
        D.: On the improved Estimation of the Compliance-Parameters
        of the Physiologically Closed Cardiovascular System, In:
        System Analysis of Biological Processes, pp. 179-187, Ed.:
        D.P.F. Möller, Vieweg Verlag, Braunschweig, 1987

[8]   Thiele, G., Möller, D., Popovic, D.: Probleme bei der
        Schätzung der Parameter eines nichtlinearen Modells des
        physiologisch geschlossenen Kardiovaskulären Systems, In
        Systemanalyse biologischer Prozesse, pp. 147-157, Ed.:
        D.P.F. Möller, Springer-Verlag Berlin-Heidelberg, 1984

A REVIEW OF RESPIRATORY SYSTEM APPLICATIONS
OF COMPUTER SIMULATION AND MODELLING TECHNIQUES

David J. Murray-Smith

Department of Electronics and Electrical Engineering

University of Glasgow, Glasgow G12 8QQ, U.K.

INTRODUCTION

There are now several different areas, within the field of respiratory system physiology and medicine, in which the techniques of mathematical modelling and computer simulation have been recognised as having special relevance in the solution of important practical problems.   Categorised in terms of physiology, these areas include the study of lung mechanics and gas flow phenomena in the airways;   the investigation of gas exchange processes within the lungs, circulation and tissues; and the study of the many highly complex processes involved in the control of breathing. However, it should be noted that the forms of mathematical model used, and the ways in which the models are developed, depend very much upon the purpose of the modelling and the ways in which the resulting models are to be applied.

Common reasons for applying mathematical modelling and computer simulation techniques to biomedical systems include:− (1) the use of models as research tools for advancing physiological and clinical insight, (2) the application of mathematical models in developing methods for the indirect estimation of physiological quantities by means of system identification and parameter estimation techniques, (3) the use of models for the development of on-line systems for control or therapy, and (4) the use of computer simulation techniques in education and training.   The aim of this chapter is to provide illustrations of these different types of application in the context of the respiratory system, using examples drawn from the areas of respiratory physiology mentioned above.

Physiological background

Within the human body most cells obtain their energy as a result of chemical reactions which involve oxygen and they must also be capable of eliminating carbon dioxide which is the end product of these oxidations. The function of the respiratory system, which comprises the lungs, the passageways leading to lungs and the chest structures responsible for movement of air into and out of the lungs, is to maintain the acidity of the extra-cellular fluid at an appropriate level by providing an adequate supply of oxygen and by removing unnecessary carbon dioxide. The gas exchange processes which occur within the body can be split conveniently into the processes of internal respiration, which involve the gas exchange between blood and body tissue, and the processes of external respiration which are concerned with the gas exchange between blood and the air within the alveoli of the lungs.

Ventilation is a cyclic process which has two main phases involving inspiration of fresh air from the atmosphere and the subsequent expiration of the alveolar air stored in the lungs. The partial pressure of oxygen is higher in alveolar air than in venous blood but there is a pressure gradient in the opposite sense for carbon dioxide. These differences of partial pressure provide the basis for pulmonary gas exchange processes and, due to the cyclic nature of ventilation, the tensions of the lung gases oscillate with the phases of the breath cycle.

Transport of oxygen between the pulmonary system and the tissues is principally achieved through the presence of haemoglobin protein within the red blood cells. Carbon dioxide is carried both in dissolved form in the plasma and in chemically combined forms. The pumping of blood by the heart propels the oxygen and carbon dioxide between the lungs and the tissues and passive diffusion induces a net movement of these gas molecules across the alveolar, capillary and cell membranes.

Rhythmical breathing is achieved by specialised neurons within the medulla which give rise to activity in the efferent nerves which innervate the diaphragm and intercostal muscles. The depth of respiration depends upon the number of active motor units, while the breathing rate depends upon the time interval between bursts of motor activity. The overall control of the system depends, in part at least, on chemoreceptors located in different parts of the body which monitor levels of oxygen and carbon dioxide.

The conceptual basis of modelling the system

Interest in mathematical modelling and computer simulation as a means of studying the elements of the human respiratory system was stimulated during the 1950's and 1960s by the availability of general purpose computing hardware capable of handling dynamic models of some complexity.   Papers describing computer simulations of the respiratory system, particularly in the context of the chemical control of breathing, started to appear with some regularity in journals concerned with biomedical engineering and in the proceedings of conferences having a control engineering or computer simulation theme.   Unfortunately, much of the earliest modelling and simulation work was not integrated with physiological experimentation and was essentially a theoretical activity performed, in the words of C. Walter, 'in a biologically sterile environment' [1].   In recent years, however, the development of mathematical models in the field of respiratory physiology has become a much more systematic and rigorous process in which experimentation and computer simulation are seen as complementary and closely related activities.

Advances in the software available for continuous system simulation have considerably reduced the development time for digital simulation programs in recent years and this has begun to have an impact on respiratory system applications. Powerful general–purpose high–level simulation languages are now available, at relatively low cost, for many different computing environments ranging from desk–top personal computers to multi–user mainframes.   Analogue simulation techniques have declined in significance but can still provide useful practical benefits in specialised applications such as those involving real–time nonlinear simulation and signal processing.

The computational facilities needed for respiratory system modelling depend almost entirely upon the nature of the application.   For example, the hardware and software facilities needed for a research environment may be quite inappropriate for a laboratory where mathematical models provide a basis for pulmonary function testing or for the development of new testing procedures.   Similarly the facilities needed for teaching and training may differ considerably from those in other application areas.

It is interesting to note that the terminology of modern computer science has begun to be applied to the processes involved in developing mathematical models [2]. Thus a 'bottom–up' modelling approach involves starting with a highly detailed system description using well established prior knowledge of the elements of the system to build up an all–embracing mathematical model which may take the form of a large and highly coupled set of nonlinear differential equations. A 'top-down' modelling process, on the other hand, involves working from a statement of the objectives towards the formulation of a simple model initially. The adequacy of this initial model is assessed by making comparisons of the model responses with corresponding measured data from the real system and modifications are made to the model, in an iterative fashion, until the model is deemed adequate for the intended application.

In the case of the respiratory system, examples of the bottom-up approach to modelling can be found in many theoretical descriptions of the processes involved in the regulation of breathing. Top-down modelling, on the other hand, has been widely used in the study of lung mechanics and pulmonary gas exchange processes, with simple models being postulated initially and additional features being added, or other changes made, on the basis of experimental evidence.

One immediate difficulty with the use of the bottom-up approach in physiological systems is that it is often difficult, or indeed impossible, to obtain reliable information on the many sub-systems within the overall model. There can also be problems in assessing the range of applicability of these models and in carrying out any form of validation since the number of variables in a typical model of this kind tends to be very much larger than the number of variables that are directly measurable in the real system.

The 'principle of parsimony' (a form of Occam's razor) states that models should be kept as simple as possible. It is important, however, to distinguish carefully between model simplification which involves demonstration that certain parts of a system have a negligible influence on the overall behaviour and model simplification for the sake only of mathematical tractability. Although the general issue of complex versus simple models in physiological systems is highly controversial [3,4] it is now generally accepted that both the top-down and bottom-up processes of model development are valuable as they serve different purposes. A useful combination of these two approaches can be achieved by splitting up a complex model so that it forms separate sub-systems which may be studied independently and assessed rigorously before being recombined within the overall model.

As models evolve they tend, almost inevitably, to become more complex as additional experimental information is incorporated. Physiological insight is important in defining appropriate boundaries for the system under study and to find some adequate way of representing any inputs which originate outside these predefined boundaries but which have important influences upon the system within. This may, in some cases, involve the use of signals from the real system recorded or stored digitally and used as variables within the simulation model. Such use of measured signals can also form part of the more general top–down modelling approach and typical non–invasive respiratory measurements which can be made in the human subject on a continuous basis include gas flow rate at the mouth, changes in body volume using a body plethysmograph, the content of inspired and expired gases and the period of the inspiratory and expiratory phases of the breathing cycle.

Model validation must be an integral part of any worthwhile modelling and simulation activity. Although no model can ever be validated in the sense that it can be proved strictly correct, one must always demonstrate that a model has a performance which is adequate for its intended application. The model can then be used until experimental evidence is produced from the real system which shows that the model is, in some respect, unsatisfactory and requires modification. In the context of research this could be taken to imply that simulation should be viewed as a negative process in that it can only be used to disprove hypotheses. Such a view of simulation is, however, a somewhat restrictive and unhelpful one and it is hoped that the examples which follow will show how simulation techniques can be applied to respiratory physiology, in conjunction with experimentation, to explain complex effects and make useful predictions that can be verified.

MODELLING AND SIMULATION AS RESEARCH TOOLS

The respiratory system is one of the areas of medical and biological research in which the use of simulation techniques is most widely accepted. This is due, in part, to the fact that the use of dynamic models based on difference or differential equations is well established in this field. In many cases simplifying assumptions had to be applied in the original formulation of these models in order to allow analytical solutions to be obtained. The present–day availability of computers and easily used simulation software has of course greatly reduced the need for closed–form solutions of model equations and has led to a re–examination and extension of many early models. Another important factor which has provided a stimulus for simulation and dynamic modelling activities in respiratory physiology is that some variables, such as the gas flow rate and concentrations at the mouth, are readily accessible for direct measurement. This means that model testing and validation is much more feasible in the case of the respiratory system than it is in many other human systems.

Mathematical modelling is often regarded as a process involving the application of physical or biological laws and principles to establish a unique set of equations to describe a given system. However, many practical applications in a research environment may also involve inverse modelling techniques in which information about the structure and parameters of the model is obtained using system identification methods from measured response data. Examples of successful applications of system identification and parameter estimation techniques to respiratory system research are to be found in a number of different areas, but particularly in the fields of gas exchange physiology and mechanics of the airways [5]. In both these areas of application the motivation has been primarily in the development of improved techniques of lung function testing and further consideration is given to applications of this type in a later section of this chapter.

The intended application always has a strong influence on the form which a model takes.
In the modelling of many engineering systems it is possible, and often desirable, to establish a close link between elements of a model and corresponding components of the real system which it represents. In the case of physiological systems, however, there are many situations where this cannot be done initially because of major areas of uncertainty within the available descriptions of the system and one objective in applying system identification may be to try to establish improved model structures for complex systems. Interesting examples which illustrate the application of identification-based methods to the problem of selecting one representation from a number of candidate models can be found in the work of Swanson on respiratory control problems [6].

Hypothesis testing is central to many research applications of mathematical modelling and simulation techniques. Results obtained through the integrated use of experimental techniques and simulation may often provide insight and evidence which would be difficult to obtain directly by conventional experimentation alone.

The study of the respiratory control system provides some interesting illustrations of ways in which simulation can be used to test hypotheses and to design improved experiments intended to throw new light on difficult physiological problems [7]. Respiratory control was, in fact, one of the first areas within physiology to which dynamic modelling and computer simulation methods were applied. In 1954 Grodins, Gray, Schroeder, Norins and Jones [8] provided one of the earliest 'chemostat' models for the chemical control of breathing. This model, which was implemented using an analogue computer, provided an accurate prediction of the ventilatory response to changes of concentration of carbon dioxide in the inspired gas mixture. Later models took account of new developments in the quantitative understanding of gas transport and exchange processes in the airways, circulation and tissues and thus provided, in control engineering terms, an improved description of the respiratory 'plant'. Examples include the descriptions of Milhorn, Benton, Ross and Guyton [9] in 1965 and of Grodins, Buell and Bart [10] in 1967. A further development by Milhorn and Brown [11] allowed the inclusion of the hypoxic drive to ventilation and attempted to partition the total ventilation between the tidal volume and respiratory rate. Milhorn, Reynolds and Holloman in their 1972 paper [12] suggested, on the basis of simulation studies and from comparisons with experimental results, that the central chemoreceptor site lay in the area between deep brain tissue and the cerebrospinal fluid.

Much of the complexity in many of these models was primarily in the respiratory 'plant' and less attention was given to the controller. Indeed, most studies have avoided many of the complications of the respiratory centre because of a lack of quantitative physiological data concerning this aspect of the system and the view, which was widely held, that within–breath (rythmic) control of breathing could be largely decoupled from the between–breath (chemical) control processes. The workings of the respiratory centre are now somewhat better understood (due in part to a model–based study by Bradley, von Euler, Marttila and Roos [13] in 1975) and, as a result, a number of respiratory control models have been produced in recent years which include cyclic ventilation and provide a means of examining competing hypotheses concerning the interaction of the neurogenic and chemical components of respiratory control [e.g. 14,15].

Although mathematical models, such as those reviewed above, can provide adequate explanations for ventilatory behaviour in response to specific input stimuli, such as carbon dioxide inhalation and hypoxia, some problems remain. One major area of difficulty is concerned with the hyperpnoea of exercise which traditional chemostat models based upon proportional control are unable to explain [16]. Essentially the problem is that for test stimuli involving exercise the error, or actuating signal, within the basic chemostat model is insufficient to account for the observed ventilation (for physiologically reasonable values of controller gain).

The problems of modelling the phenomena associated with exercise hyperpnoea were reviewed by Grodins in 1964 [17] and again in 1981 [18]. Other controller structures have been suggested and much experimental effort has been devoted to the study of arterial oscillations of CO_2 and O_2 tension and their effects, in terms of timing, on both the carotid chemoreceptor discharges and the breath–by–breath pattern of ventilation. The evidence now available suggests that these oscillations could, indeed, contain the information necessary to establish the rate of CO_2 output. The peripheral chemoreceptors can certainly detect such oscillations and, in principle, this could allow the control system, with a feedforward type of controller structure, to compensate for the effects of exercise. Whether or not the system uses such information in this way is still not known. Detailed accounts of some of the physiological and modelling aspects of this problem have been provided by Petersen and Cunningham [16], Herczynski [19] and Murray–Smith and Carson [20,21] in a recent volume on the respiratory system edited by Cramp and Carson [7]. The evidence at present available suggests strongly that it is only through a systematic process of model extension and testing against experimental results that an acceptable theory of human respiratory control can evolve.

While the discussion of exercise hyperpnoea illustrates the fact that some fundamental physiological problems remain unsolved, certain aspects of earlier work involving respiratory control system models are now receiving attention in the context of clinical research. One point, which becomes clear from any review of the literature on modelling and simulation work involving respiratory control processes, is that even relatively simple mathematical models of the respiratory system can display self–sustained oscillatory phenomena and thus exhibit periodic patterns of breathing of a type similar to those observed clinically in conditions such as Cheyne–Stokes breathing [e.g. 22]. Early examples of modelling and simulation work involving the study of periodic breathing phenomena include contributions by Milhorn and Guyton [23], Longobardo, Cherniack and Fishman [24], Lange and others [25] and Cherniack and Longobardo [26]

More recently Khoo and his colleagues [27] developed a general model to explain periodic breathing under a variety of normal and clincial conditions. The model structure includes separate compartments for the lungs, body tissue and brain tissue and incorporates representations of both the central and peripheral chemoreceptors. Under normal physiological conditions this model can exhibit periodic breathing if the loop gain is increased, and the model has been used successfully to investigate periodic breathing phenomena in adults. Nugent and Finley [28] have also applied a version of this model to investigate problems of periodic breathing in infants. The results of this model–based study agree with clinical observations in that under normal breathing conditions the model of the respiratory control system is relatively stable. However, certain combinations of parameter values and levels of arterial O_2 and CO_2 lead to situations in which the stability is marginal and periodic breathing could result.   In all the cases studied changes in the content of the inspired gas mixture could be introduced to make the system more stable, which is again consistent with clinical observations.

Many other examples could be provided to illustrate current trends in the use of modelling and simulation techniques in respiratory system research.   Case studies can be found in the physiological and bio-medical engineering literature as well as in two texts dealing specifically with the mathematical modelling of the respiratory system [7,29].   Papers published within the 'Modeling Methology Forum' of the American Physiological Society in the 'American Journal of Physiology' and 'Journal of Applied Physiology' provide other useful illustrations of modelling principles.

SIMULATION FOR TEACHING

In attempting to gain a proper understanding of respiratory physiology students are faced with nonlinear dynamic systems of considerable complexity.   Even with a restricted field, such as the study of pulmonary gas exchange, there are very considerable problems in terms of the number of processes involved and the number of separate factors which are important.

Models which have been developed to assist in the teaching of respiratory physiology and associated clinical subjects, such as anaesthesia, are now   quite numerous and three typical examples will be considered.   The first is a simple model of carbon dioxide transport which is intended to illustrate concepts of pulmonary gas exchange which present difficulties in terms of conventional teaching methods.   The second example is the 'Macpuf' model which is widely used in clinical teaching and which attempts to simulate the lungs, airways, circulation, tissues and control of breathing for a wide range of normal and abnormal conditions.   This model is used in preclinical and clinical courses and in the professional training of anaesthetists. The third case study involving teaching is concerned with a real-time computer simulation used for general anaesthesia training.

91

Simulation in respiratory gas exchange teaching

Students being introduced to the concepts of respiratory gas exchange are faced with the task of gaining an understanding of a complex system which involves a large number of important variables, including ventilation, perfusion and the concentrations of respired gases in the inspired air, in the alveoli, in the blood and in the tissues. The system, as well as being complex, is nonlinear and students may experience considerable difficulty in grasping basic concepts, particularly in terms of the interaction of ventilation and perfusion. Under abnormal or diseased conditions the problems of interpreting the mechanisms of gas exchange become even greater.

A mathematical model involving a time varying representation of ventilation has provided the basis for analogue and digital computer simulations which can augment conventional methods of teaching the concepts of pulmonary gas exchange [30]. The model, which is restricted to the study of carbon dioxide, involves only three lung compartments. These are the dead space (volume V_D) and two alveolar compartments (volumes V_{A1} and V_{A2} having blood flows $\dot{Q}_1$ and $\dot{Q}_2$ respectively). There is also a tissue compartment which, if required, may be omitted from the model and mixed venous blood with a constant pre–determined partial pressure of carbon dioxide can flow to the lungs. The structure, variables and principal parameters of this model are shown in Figure 1. If the tissue compartment is included within the model a variable metabolic CO_2 production $\dot{M}$ and an initial condition for the CO_2 tissue tension P_{TC} must be introduced.

In terms of this model structure, which is based upon the work of Murphy [31,32], the respiratory cycle consists of three stages:–

Stage (1) Inspiration of dead space gas. This stage is governed by the condition

$$\int_0^t \dot{V} \, dt \leqslant V_D \quad \text{and} \quad \dot{V} \geqslant 0$$

Stage (2) Inspiration of atmospheric air.

$$\text{Here} \int_0^t \dot{V} \, dt > V_D \quad \text{and} \quad \dot{V} \geqslant 0$$

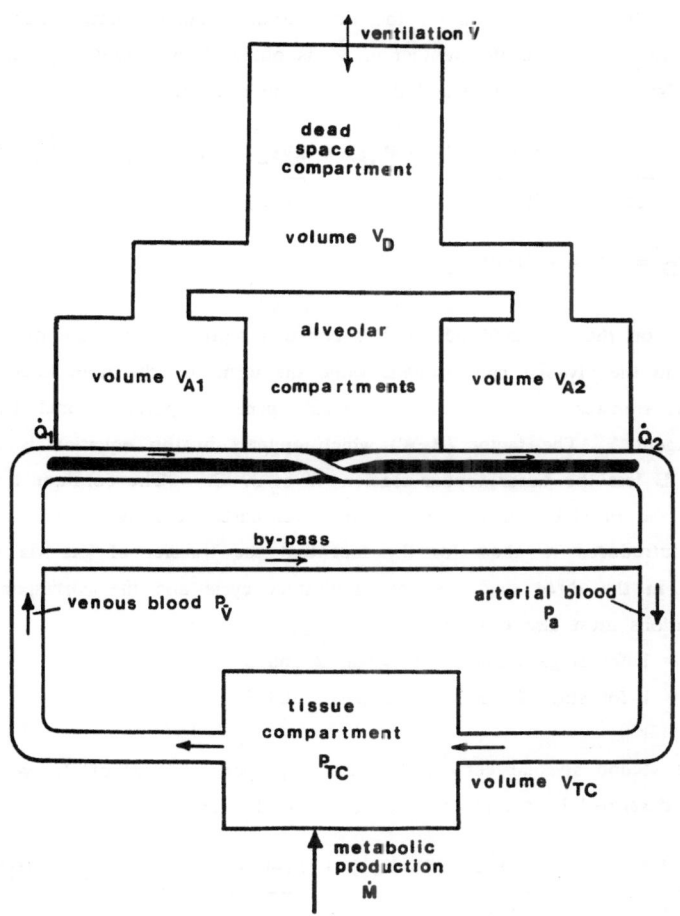

Stage (3) Expiration.　For this stage

$$\dot{V} < 0$$

The differential equation for the first alveolar compartment, which is assumed to have a proportion k of the ventilation $\dot{V}$, is obtained by considering the mass transfer of CO_2 from the dead space and tissues.   The equation is

$$\frac{V_{A1}}{(B-W)} \frac{dP_{A1}}{dt} = \frac{k\dot{V}S_1}{(B-W)} (P_D - P_{A1}) + \frac{k\dot{V}S_2}{(B-W)} (P_I - P_{A1}) + \dot{Q}_1 b(P_{TC} - P_{A1})$$

where $P_D = kP_{A1} + (1-k)P_{A2}$

The term on the left hand side of this equation represents the rate of change of mass of CO_2 in the alveolar compartment while the terms on the right hand side represent the mass transfer of CO_2 from the dead space compartment and from the tissue compartment.   The factor (B-W) which appears in this equation is the barometric pressure B less the water vapour tension W at 37°C.   The constant b represents the slope of the physiological dissociation curve for carbon dioxide.   The symbols S_1 and S_2 are introduced to allow for the fact that the transfer of gas via the airways is different in the three stages of the respiratory cycle and the corresponding terms in the equations must also change for each stage.   Hence

$S_1 = 1$ for stage 1 and 0 for stages 2 and 3

$S_2 = 1$ for stage 2 and 0 for stages 1 and 3

The second alveolar compartment has a proportion (1-k) of the ventilation and is similarly described by a differential equation of the form.

$$\frac{V_{A2}}{(B-W)} \frac{dP_{A2}}{dt} = \frac{(1-k)\dot{V}S_1}{(B-W)} (P_D-P_{A2}) + \frac{(1-k)\dot{V}S_2}{(B-W)} (P_I - P_{A2}) + \dot{Q}_2 b(P_{TC} - P_{A2})$$

The third equation represents conditions in the tissue compartment (volume V_{TC}) and simply states that the rate of change of mass of CO_2 in the tissue compartment is equal to the difference between the metabolic production of CO_2 and the transfer of CO_2 to the two alveolar compartments.   That is

$$\frac{bV_{TC}}{dt} \frac{dP_{TC}}{} = \dot{M} - \dot{Q}_1 b(P_{TC} - P_{A1}) - \dot{Q}_2 b(P_{TC} - P_{A2})$$

The equations of this low order lumped-parameter model have also formed the basis of research investigations involving the use of system identification techniques for the development of non-invasive methods for the estimation of cardio-pulmonary quantities [33-37] and a similar form of model has been used successfully to represent the cyclic gas exchange process within a much larger model of the respiratory control system [15].

For teaching uses the model was implemented using both analogue and digital simulation techniques [30]. The analogue simulation was developed in the form of a special purpose desk−top simulator constructed using readily available low cost components to achieve an overall level of computing inaccuracy of 1% or less. One of the objectives in the design of the simulator was to provide a convenient and easily understood interface for the user with all controls easily identified and calibrated in physiological units. The use of the simulator required no knowledge of analogue computing techniques. Time scaling facilities were included so that the equations could be solved at a rate faster than real time to provide rapid feedback to the user about the effects of parametric changes on the model behaviour. Two outputs were provided for graphical display with a choice of CO_2 tension in each of the alveolar compartments, mean alveolar CO_2 tension, tissue tension and mean arterial tension as variables to be selected. Appropriate zero level and calibrating signals could also be output to provide reference levels for interpretation of the graphical records.

The digital simulation was implemented, for reasons of portability, as a FORTRAN program. The student was provided with information in the initial part of the program about the parameters that could be altered and the variables to be output. During the simulation the student could interrupt the execution of the program and change either a parameter value or an output variable. As well as graphical output a listing of all the appropriate variables and the set of model parameters which was used was provided for the student when the simulation run was terminated. In addition the program calculated results which would be obtained from the application of certain pulmonary function tests to the simulated lung. These tests results were provided to allow the student to relate the findings from standard clinical tests to the underlying physiology of gas exchange processes.

For both the analogue and digital implementations two important educational objectives were the demonstration of the relationship between alveolar ventilation and perfusion in the determination of alveolar gas tension, and the mechanisms through which an arterial−alveolar tension difference is created. Another useful feature of this model is the facility to show the significance of dead space in relation to alveolar ventilation and alveolar gas tension levels. The possibility of demonstrating the time course of changes in alveolar and mixed venous tensions following a change in ventilation and of showing the magnitude of the variation in alveolar tension during the respiratory cycle are other important benefits.

The 'MacPuf' model [38]

The use of specialised sub-system simulations, such as that outlined above for the gas exchange process, is becoming relatively commonplace in the teaching of physiology. An alternative approach involves the development of much larger and more complex computer models which incorporate many sub-models and can therefore be used to illustrate a large number of physiological principles. The penalty in larger models of this kind is that any implementation as a simulation using general-purpose digital computers must involve execution times which increase with the model size. This time factor can present a significant problem in applying large and complex models in teaching situations since interactive use of a simulation becomes quite inappropriate if the time required for solutions exceeds the attention span of the student. Ideally, a teaching simulation should be capable of being run in a number of different time scales in order to allow students to study both the short term behaviour of subsystem models, such as the gas exchange processes where time constants may be of the order of seconds, and the longer term behaviour of the complete system which may involve controlled processes which are very slow in comparison. For example a student might wish to simulate deep-sea diving where the diver is exposed to high pressures for periods of many minutes, or even hours, and to use the simulation to investigate different decompression strategies. This problem could be very time consuming, especially if the frame time used in the simulation of these long term processes is the same as that used in the investigation of shorter term phenomena within subsystems.

One interesting example of a complex respiratory system simulation widely used for teaching is 'MacPuf' [38]. This was developed over a six year period by C.J. Dickinson and his colleagues at McMaster University, University College Medical School and St. Bartholomew's Hospital Medical College. Although intended originally to be used exclusively for teaching, this digital simulation has found a variety of other applications in clinical work and in research. The computer program, which was written in FORTRAN, was made available, at various different stages of its development to other medical schools in Europe and North America and has been adapted and extended by others to suit their own particular needs.

The main physiological structures and functions which are incorporated within the basic MacPuf simulation include blood circulation, gas exchange, the control of ventilation and tissue metabolism. The computer program allows the operator to make many different changes including, for example, inspired gas concentrations, cardiac performance, metabolic rate, haemoglobin concentration, whole body

96

bicarbonate, lung volume, body temperature, barometric pressure and lung elastance. Alterations of these and of other parameters can produce acute changes in the regulatory mechanisms of the model.   In many cases a new steady state condition will be achieved after the initial transient which follows the parameter change, but situations do arise in which no steady state can exist.   In such instances some form of sustained oscillation may be observed, as in Cheyne–Stokes breathing, or the values of variables within the model may pass beyond the range consistent with life being maintained.

The 'MacPuf' model includes facilities for pulmonary function testing and allows for the collection of expired gases for analysis.   Bag rebreathing experiments and artificial ventilation can also be included when required.   Clinical disorders can be simulated by establishing appropriate conditions on the states and parameters of the model.   Conditions such as asphyxia, ventilatory failure, hypothermia and cardiac arrest provide examples of common abnormal situations which can be handled readily within the simulation.   Possible therapeutic measures can be introduced by the student and their effects, beneficial or otherwise, observed.

The most important feature of 'MacPuf', which distinguishes it from many other computer–based models of respiratory processes, is that provides a reasonably accurate representation of many different aspects of the system within a single simulation program which runs in an interactive mode in a time scale which is convenient for teaching applications.   This combination of a complex model structure and a relatively high speed of execution in comparison with the dynamics of the real system was achieved by simplifying the description, as far as possible, at each stage with the aim of achieving an accuracy in the simulation consistent with the physiological uncertainties.   Greater accuracy is of no value, especially if it can be achieved only at the expense of execution time.

The most significant measure adopted in 'MacPuf' to ensure that a high speed could be achieved was the choice of frame times of the order of 5 to 10 seconds. This was very much longer than the sampling interval used in previous respiratory system simulations incorporating similar subsystem detail and was introduced simply to make the model usable on a small computer.   Special numerical techniques were used to introduce additional damping and thus give the numerical stability needed to ensure that the simulation would behave in a physiologically meaningful way with this long sampling interval.

The above description shows that 'MacPuf' differs very considerably from the gas exchange teaching model which was outlined previously. The digital simulation for the gas exchange model involved many sample periods within each breath cycle which, typically, would be of four seconds duration. The objectives in the design of the two simulation programs are, however, very different. The 'MacPuf' simulation is an implementation of a comprehensive model which allows the behaviour of the complete respiratory system to be compared with clinical and physiological observations over time intervals which may be quite long. The gas exchange model, on the other hand, allows more detailed consideration than is possible with 'MacPuf' of one of the sub–systems involved in respiration and is of value for the study of events having a relatively short time course.

A computer simulation for training in anaesthesia

The importance of ensuring that trainee anaesthetists are given experience in dealing with the type of anaesthetic emergencies that are likely to occur from time to time in any operating theatre has led recently to increased interest in simulators designed specifically for anaesthesia training. Such training simulators can, in many respects, be likened to the flight simulators which for many years have been used, with great success, in the training of airline pilots. The representation of oxygen and carbon dioxide transport processes is an important element of any simulation developed for the study of inhalational anaesthesia and such simulations are therefore closely linked to respiratory system models. The gas transport model must, however, include predictions of airway pressure and flows as well as the exchange of gas in the lungs and tissues. The model must also include, where necessary, representations of any external hardware, such as ventilators and bags, used by the anaesthetist.

A simulator developed recently by Schwid [39] provides an interesting example of the way in which standard general–purpose computer hardware can provide a valuable training facility. The objectives in the development of this simulator were to provide support for teaching the essential concepts of anaesthesia, to offer a safe and convenient way of training for emergency situations and to provide a means of improving diagnosis and treatment of unusual, and possibly life–threatening, complications.

The development of simulators for anaesthesia training involves two main problem areas. Firstly, it is essential to have a computer model which can be run in real time, and can predict the response of different patients for a wide range of physiological states to a variety of drugs. Such a model must allow for drug interactions and must be capable of predicting, accurately, the responses of different populations of patient to these drugs. The second important problem area in the development of such simulators is in the design of a suitable user interface. The more expensive options could involve a mock-up of the operating theatre and a mannequin to represent the patient, together with an anaesthetic machine and all the monitors likely to be available in the real situation. Less expensive options might be based on the use of computer graphics but inevitably there is some loss of realism in using such an interface.

Schwid's simulation runs on a DEC VAX 11/730 computer with the programming carried out using FORTRAN. The simulation involves three main sub-models and uses the multiple model approach developed by Beneken and his colleagues [40,41]. Of these three sub-models one represents oxygen and carbon dioxide transport, the second is a model of the circulation and the baroreflexes while the third involves pharmacokinetics and pharmacodynamics. The circulatory system sub-model predicts cardiac output and the systematic arterial, central venous, pulmonary artery and left ventricular pressures. The gas transport sub-model involves a representation of the pulmonary gas exchange processes which is based upon the simple three compartment model of Riley and Cournand [42]. There are also eight tissue compartments, each of which is defined in terms of mass, oxygen consumption and carbon dioxide production, and can be associated with specific body organs. The pharmacokinetic model describes the tissue uptake of drugs, including both intravenous and inhalation agents.

Although described above as independent sub-models, it is important to note that these elements of the simulation interact in a number of different ways. For example, the circulation clearly determines the level of tissue perfusion and thus has a direct effect on the gas transport model. Equally important, however, is the fact that the tissue oxygen and carbon dioxide tensions can influence circulatory parameters. Similarly, ventilation determines the uptake of inhalation agents and thus influences events within the pharmacokinetic and pharmacodynamic model; however the resulting drug levels can in turn affect the metabolic processes in the tissues and can thus have a direct effect on the gas transport sub-model. The model is therefore large and complex but has been extensively tested both at the sub-model level and as a complete simulation.

The interface developed by Schwid consists of a mouse-driven input and a graphics display output on a modified Silicon Graphics IRIS workstation. The patient is represented graphically and responds to the actions of the trainee anaesthetist by changing pupil size and skin colour or by closing the eyes.   The display includes items such as an anaesthetic machine, an oxygen analyser and representations of other appropriate hardware as well as standard monitors such as the electrocardiogram and pressure waveforms.

The initial design and construction of this simulator has been completed and extensive model validation tests have been successfully carried out.   Together the model and the display system should be capable of providing a powerful simulator for general anaesthesia training.   As the costs of computer hardware fall it will be very interesting to see how long it will be before simulators become common aids for the training of anaesthetists.   In the case of airline pilots the economic benefits of using simulators for training are well established.   For anaesthetists, however, the benefits of improved training are most likely to be seen in terms of a reduction of mortality figures and, as in many clinical situations, any cost–benefit analysis presents considerable difficulties.   It is clear, however, that the use of simulation techniques in general medical teaching is increasing rapidly.   The type of interactive simulation typified by the small gas exchange model and by the much larger 'MacPuf' type of model, has been found already to produce significant benefits in clinical and physiological teaching.   Further work is, however, clearly necessary in order to make the best use of powerful educational resources such as these.

ESTIMATION TECHNIQUES INVOLVING SYSTEM IDENTIFICATION METHODS

System identification, which has already been discussed briefly in the context of research applications, can provide a basis for non–invasive methods for the estimation of physiological quantities which have direct clinical importance.   The use of computers as diagnostic aids is a closely associated topic for which model–based identification techniques have obvious relevance.   System identification methods are also of importance for on–line control, therapy and drug administration.

A number of general review papers have been published on the application of identification and system parameter estimation techniques to biological and physiological systems [e.g. 43–46].   An additional review by Linkens [5] is concerned specifically with identification studies of the respiratory and cardiovascular systems.

Applications involving gas exchange models

Although many different models are available which describe aspects of pulmonary gas exchange and gas transport processes, a large proportion of these are of very limited value in the context of system identification in that they are based upon steady-state assumptions.   These steady state models have been reviewed by Rahn and Fenn [47] and by Otis [48].   The use of dynamic models provides much more flexibility in the design of experiments to provide response data which is rich in information for system identification and parameter estimation purposes.

In 1977 Murray-Smith and Pack [49] published a review of dynamic gas exchange models which included discussion of the application of these models to the estimation of cardio-pulmonary quantities.   Much of the interest in the use of parameter estimation in asociation with dynamic models of gas exchange processes arises from the possible measurement of the total pulmonary blood flow, or cardiac output, from gas flow and concentration measurements at the mouth.

Earlier published research on the estimation of cardiac output as a parameter within a gas exchange model included the work of Bekey and Maloney [50,51] which was based upon an adaptation of a continuous gas flow model of Grodins [8,10]. The use of a model incorporating cyclic ventilation has led to descriptions similar in many respects to the gas exchange model already discussed in the section of this chapter concerned with teaching applications.   In most cases, however, the structure used has involved a homogeneous lung having only one alveolar compartment.   In some instances [e.g. 36] the work has been concerned not only with the use of parameter estimation for the indirect measurement of specific quantities, such as cardiac output, but also with investigation of the potential of these techniques for the development of new respiratory function tests which do not depend upon steady state conditions of ventilation.

A paper by Bache and others [36] provides an interesting example of the application of system identification techniques to a homogeneous lung model and shows some of the benefits which can result from an integrated approach to the theoretical and experimental aspects of the work in which simulation plays an important part. The gas exchange processes of the lungs and tissues in the normal human subject were represented by the lumped parameter compartmental model of Figure 2.   The model, which comprised a rigid dead-space compartment, a single compliant alveolar compartment and a single tissue compartment, represented only the CO_2 gas exchange processes and incorporated a linearised form of the CO_2 dissociation curve.   The model output was taken to be the partial pressure of CO_2 in the alveolar compartment ($P_A(t)$) which could be compared, for system identification purposes, with end–tidal measurements at the mouth. Test stimuli were applied through variation of the measured ventilation at the mouth and by imposed changes of partial pressure of the inspired gas ($P_I(t)$).

Structural identifiability investigations based upon the approach of Bellman and Åström [52] showed that, in principle, four parameters ($\dot{Q}$, V_A, V_{TC} and $\dot{M}$) could be estimated independently, together with the initial values of partial pressure in the alveolar and tissue compartments, provided the slope and intercept were known for the linear representation of the CO_2 dissociation curve. Numerical identifiability was investigated using standard measures based upon the parameter information matrix and parameter correlation matrix [53,54]. This showed that simple forms of test input of the type used in previous studies [e.g. 35] (such as a step change of carbon dioxide partial pressure in the inspired mixture) could produce situations of near unidentifiability [36,37]. The use of persistently exciting forms of test signal was therefore assessed through simulation methods in conjunction with the theory of optimal test signal design. A form of persistently exciting input was found which could be applied by manual operation of a three–way valve and thus facilitated routine clinical application. This input took the form of a square waveform for $P_I(t)$ with equal intervals of air and CO_2 breathing and was found to give significantly improved results when compared with parameter estimates obtained from simple step tests [36,37]. Constraints relating to the validity of the mathematical model, and the assumptions upon which it was based, limited the duration of experiments to a maximum of ten minutes with input CO_2 concentrations of up to seven percent. For each of the unknown parameters an optimal switching frequency for the square waveform was then determined using the D–optimal criterion [55]and other related test signal design criteria [53].

Simulation studies showed that the optimum test signal design for estimation of the cardiac output parameter ($\dot{Q}$) was significantly different from the optimum designs for other parameters taken one at a time. The optimum design for the case where equal weighting was given to all of the parameters was also found to be significantly different and involved a switching period approximately double that of the best design for estimation of $\dot{Q}$ alone. Test signals with a relatively high switching frequency were found to be necessary for the accurate estimation of V_A whereas much lower frequencies were optimal for the parameters $\dot{M}$ and V_{TC}. This makes sense in terms of the physiology of the system since the parameter V_A is associated with the small time constant of the alveolar compartment while $\dot{M}$ and V_{TC} are paremeters of the tissue compartment which has much slower dynamics. The cardiac output parameter, $\dot{Q}$, is associated with the dynamics of gas transport between the tissue compartment and the alveolar compartment and it could therefore be expected that a switching period for accurate estimation of this parameter might lie, as in fact it does, between the values for V_A and V_{TC}.

The predictions resulting from the simulation studies were assessed experimentally in a series of tests involving human subjects whose lungs were 'normal' in terms of routine pulmonary
function tests and for whom the structure of Figure 2 was believed to be appropriate [36]. The results showed that very significant practical benefits were obtained from the use of the 'optimal' test input signals and experience with the application of this identification—based approach to the non—invasive estimation of cardio—pulmonary quantities has been encouraging both in terms of the reproducibility of estimates and of comparisons with other methods of measurements [36,37].

Other work concerned with the identification of gas exchange parameters includes that of Wiberg and colleagues who have successfully used an extended Kalman filter as a pulmonary blood flow indicator [56], and for the estimation of lung capacity and dead space [57]. The same group has also carried out critical studies of the effects of modelling simplifications on estimates of pulmonary perfusion when multiple soluble gases are employed [58].

Single—breath methods for the estimation of pulmonary blood flow have received considerable attention since their introduction by Kim, Rahn and Farhi [59] in 1966. The method does not involve any foreign inert gas or special respiratory manoeuvres such as breath holding or rebreathing from an external bag. In principle, all that is required is a set of measurements of gas concentration and flow at the mouth during a single prolonged expiration. A simple homogeneous lung model then provides the basis for the calculation of pulmonary blood flow.

In spite of its inherent simplicity, the single breath method has not gained widespread acceptance largely because its accuracy is questionable and attempts to evaluate the method experimentally through comparisons with other techniques have produced contradictory results [60]. Errors appear to arise both as a result of the modelling assumptions and the data analysis techniques [61].

Rebreathing techniques can also provide a basis for the estimation of parameters of gas exchange models [e.g. 62]. The factors which limit the reproducibility and accuracy of results currently obtained by this approach are again not fully understood although the method, like the single–breath approach, has been quite widely investigated. Recent analysis by Weisiger and Swanson [63] suggests that a method of data analysis based upon a mathematical model involving cyclic ventilation might allow

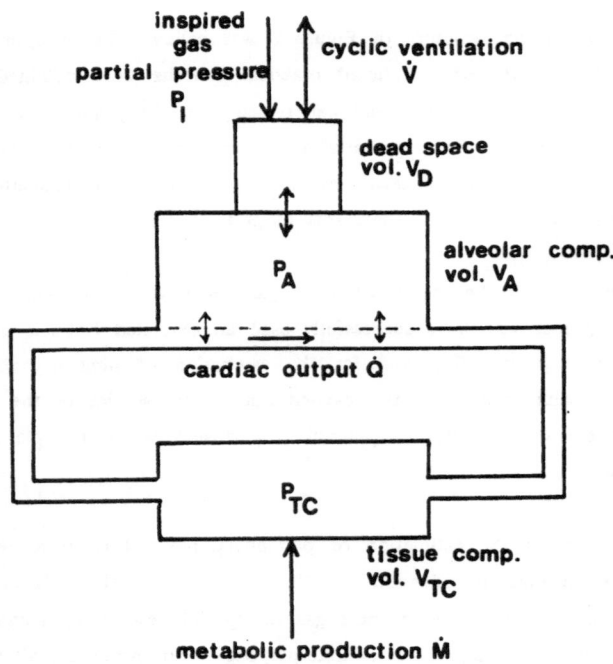

Figure 2   The structure, variables and parameters of
             the homogeneous lung model.

improved estimates to be obtained from rebreathing data.

Other recent work involving modelling approaches to the estimation of clinically important parameters of the gas exchanging processes include the research of Prisk and McKinnon [64] on the estimation of carbon monoxide diffusing capacity and their work on estimation of other parameters from carbon monoxide uptake measurements [65]. The investigation of multiple inert gas elimination measurements as a basis for pulmonary function testing has attracted interest for many years and both analytical and simulation–based studies continue more to throw light on this complex subject [e.g. 66].

Lung mechanics models

In the respiratory field much interest has been shown in the application of parameter estimation techniques to mathematical models of the airways. Inspired air must reach the alveoli from the mouth by flow through a complex branching structure of airways each of which may be characterised by parameters such as resistance and compliance. Lumped parameter models have for long been used to describe this system and a typical representation, which is relevant in clinical terms, involves an equation

$$P = P_m - P_{p\ell} = R_1\dot{V} + R_2\dot{V}^2 + EV$$

where P_m is the mouth pressure, $P_{p\ell}$ is the pleural pressure, $\dot{V}$ is the velocity of air flow and V is the tidal volume. The constants R_1 and R_2 are associated with the linear and nonlinear resistive effects and E is the elastance of the lung and chest wall. The term involving R_2 allows for turbulent flow and is especially important during forced expiration or for normal expiration with patients suffering from chronic obstructive lung disorders.

Methods which have been widely used to estimate parameters of this equation from measurements include some based upon instantaneous sampling of the airflow, volume and transpulmonary pressure [e.g. 67], others involving extensions of the loop–flattening technique of Neergaard and Wirz [68], or body plethysmography [69]. A critical review of these techniques has been prepared by Nada [70] who states that all of these technqiues have practical limitations caused either by problems encountered in their use with patients suffering from respiratory disorders or due to computational problems which arise when noise is present in the measured data. In some cases the available evidence suggests that when R_2 is neglected biased estimates may be obtained for the parameters R_1 and E.

An alternative approach has been proposed by Nada and Linkens [71] which involves an adaptive tracking method based upon a steepest descent algorithm. The technique is based upon adaptive control theory and has been implemented both in analogue and digital form [72]. The digital tracking system has been applied successfully for the estimation of lung parameters from neonatal data. Both the traditional and adaptive tracking methods are discussed by Linkens in his 1985 review paper [5].

A different approach to the investigation of lung mechanics is provided by the use of forced oscillation techniques to determine impedance. This allows investigation of the frequency dependence of flow resistance [73]. Spectral analysis techniques have been found to provide significant benefits in this context, in that complex signals can be used to provide simultaneous estimates of impedance over a range of different frequencies [74,75]. Spectral methods also allow coherence measures to be used to investigate the validity of linear and nonlinear models relating pressure and flow.

MODELS FOR ON-LINE CONTROL

Interest in models of the respiratory system for on-line control has arisen largely from the needs of those who design equipment for artificial ventilation and anaesthesia. One example is to be found in the work of Baker and Hahn [76] who used simulation techniques, in conjunction with a lung mechanics model, to investigate the effects of alteration in the inspiratory time and gas flow waveform in artificial ventilation. Investigations of alveolar and pleural pressures and inspiratory work allowed a pattern to be found which caused the least physiological upset. Another paper, by Epstein and Epstein [77], reports on the development of models which allow the prediction of the flow of gas delivered to the lungs of infants when subjected to mechanical ventilation.

In the anaesthetics field interesting developments are taking place in terms of automatic systems for the control of inhalational anaesthesia. Tatnall [78] has described one such system which combines clinical measurement with a respiratory model, the parameters of which are used to establish the control algorithm for the closed-loop system. The resulting control system requires no prior information concerning patient variables and the only quantitative information to be provided by the anaesthetist is the desired level of gas concentration at the alveoli and the maximum permitted inspired concentration. The patient characteristics are identified as anaesthesia proceeds through parameter estimation and patient state tracking.

106

Research in the field of respiratory control has led to a requirement for automatic control systems which regulate the level of alveolar oxygen or carbon dioxide at a set level. Such systems can also be used to provide forcing functions for system identification purposes.

A system for control of the partial pressure of inspired carbon dioxide was described by Lambertson and Wendel [79] in 1960. Systems of a similar kind have been further developed by Bellville and others [80] and by Swanson and his colleagues [81] using an on-line hybrid computing system. Swanson and Bellville [82] used their system to manipulate the inspired gas concentrations to achieve an end-tidal forcing function which caused the end-tidal gas concentrations to follow specific waveforms. These waveforms acted as the test input forcing function for the respiratory controller in the human subject under test while the ventilation provided a measure of the corresponding respiratory controller output.

Another implementation of a system of this kind has been described by Chambille and others [83]. This involved a system which could independently control either the alveolar partial pressure of oxygen or of carbon dioxide during both steady states and transients. Like the other systems this 'alveostat' was developed as a research tool for the investigation of respiratory control processes.

Success in the design of any closed-loop system depends, in part, upon the designer's understanding of the characteristics of the system being controlled. The controlled 'plant' in the case of these on-line systems involves the gas exchanging processes of the lungs. Since quantities such as tidal volume and alveolar ventilation are not directly measurable some model-based processing must be carried out within these systems. In many cases this has involved the use of analogue computer hardware within the closed-loop system, providing interesting examples of 'hardware-in-the-loop' simulation in a biomedical context. The accuracy of control clearly depends upon the validity of the plant model which provided the basis for the control system design and for the implementation in terms of on-line signal processing.

DISCUSSION

The mathematical modelling of physiological systems is an iterative process which should involve the following:-

1) Careful consideration of the modelling objectives.

2) Careful experimental deisgn to obtain data for identification or model testing purposes.

3) Definition of the model structure and estimation of model parameters from measured response data as well as from *a priori* information.

4) Extensive testing of the model to determine whether or not it is consistent with all of the available experimental data from the real system.

If tests on a model, either using data applied at the identification stage or other measured data sets, are unsatisfactory or show inconsistencies, then that model cannot be applied until modifications have been made and the resulting model re-tested. If all tests on a model are satisfactory it may then be used for its intended application until new evidence suggests some further need for changes.

Interest in the application of system identification and parameter estimation techniques as part of an integrated experimental and theoretical approach to the modelling of complex systems is growing rapidly at the present time [84]. Computer simulation techniques have a very important role within an integrated approach of this kind since, even when based upon incomplete and inaccurate models, they can be of great value for the design of new experiments and in suggesting additional measurements which could be useful for the further development and improvement of the model. Simulated response data can be particularly valuable for the development and verification of new software tools for system identification and parameter estimation. Modelling and simulation thus represent one aspect of what might be termed a 'systems approach' to problem solving.

There is an important distinction in the way in which mathematical modelling is used by physiologists and the way in which it is employed by engineers. Engineering models are usually well understood, although possibly quite complex, and are developed as a means of carrying out simulation experiments which would otherwise be impossible because of cost or safety. In the modelling of the respiratory system, as in other physiological areas, the objective is usually quite different. Most respiratory models are developed either as a stepping stone in research, aimed primarily at providing a better understanding of the real system, or for use in new methods of indirect measurement in fields such as pulmonary function testing. Models used in teaching are also usually based upon models developed initially in the course of research.

In both research and teaching complexity provides a compelling reason for attempting to build formal mathematical models of the system. Computer models can readily expose any weakness inherent in verbal descriptions and may, on occasions, reveal unexpected explanations. The real test of a model is that it not only shows quantitative agreement with the physiological system but that it also can suggest new experiments and provide new insight. The physiological literature is full of important information gaps which only become apparent when a rigorous systems–based approach is applied.

The examples which have been mentioned in this chapter provide only a small sample of the mathematical modelling and computer simulation activities in the field of respiratory physiology and medicine. It is hoped, however, that they illustrate the variety of modelling approaches which are relevant and show something of the range of their applications in physiological and medical research, in teaching and in patient care.

Future developments in this field will depend in part on the provision of easily used software to ensure efficient and effective interaction between the modeller, the experimental data and the computer itself. For example, better facilities are needed to allow simulation models to be used in conjunction with large databases involving experimental records.

Ideally one wants the performance of the computer to be well matched to that of the user for all of the complex iterative processes of model development, testing, evaluation and restructuring which are so essential in a properly integrated systems approach. Developments in the field of artificial intelligence may well contribute to the provision of an integrated modelling environment through expert systems incorporating experimental databases and knowledge bases specific to the respiratory field.

REFERENCES

[1] Walter, C., Preface, in Solomon, D.L. and Walter, C. (eds.), *Mathematical Models in Biological Discovery*, (Lecture Notes in Biomathematics No. 13), Springer Verlag, Berlin, 1977.

[2] Cobelli, C., 'Identification of endocrine–metabolic and pharmacokinetic systems', in Barker, H.A. and Young, P.C. (eds.), *Identification and System Parameter Estimation, 1985*, Pergamon Press, Oxford, 45-54, 1985.

[3] Garfinkel, D., 'Complex biological models: their construction and effective use', *Math. Comput. Simul.*, XXIV, 425–429, 1982.

[4] Guyton, A.C., 'On the value of large models of biological systems', *J. Cyb. Inf. Sci.*, 2, 71–72, 1979.

[5] Linkens, D.A., 'Identification of respiratory and cardiovascular systems', in Barker, H.A. and Young, P.C. (eds.), *Identification and System Parameter Estimation, 1985*, Pergamon Press, Oxford, 55–67, 1985.

[6] Swanson, G.D., *Dynamic Forcing in the Study of the Human Respiratory System*, Ph.D. Thesis, Stanford University, Stanford, California, 1972.

[7] Cramp, D.G. and Carson, E.R. (eds.), *The Respiratory System*, (Measurement in Medicine Series, Vol. 2), Croom Helm, London, 1988.

[8] Grodins, F.S., Gray, J.S., Schroeder, K.R., Norins, A.L. and Jones, R.W., 'Respiratory responses to CO_2 inhalation.   A theoretical study of non–linear biological regulators', *J. Appl. Physiol.*, 7, 283–308, 1954.

[9] Milhorn, H.T., Jr., Benton, H., Ross, R. and Guyton, A.C., 'A mathematical model of the human respiratory control system', *Biophys. J.*, 5, 27–46, 1965.

[10] Grodins, F.S., Buell, J. and Bart, A.J., 'A mathematical analysis and digital computer simulation of the respiratory control system', *J. Appl. Physiol.*, 22, 260–276, 1967.

[11] Milhorn, H.T., Jr., and Brown, D.R., 'Steady–state simulation of the human respiratory system', *Comput. Biomed. Res.*, 3, 604–619, 1971.

[12] Milhorn, H.T., Jr., Reynolds, W.J. and Holloman, G.H., Jr., 'Digital simulation of the ventilatory response to CO_2 inhalation and CSF perfusion', *Comput. Biomed. Res.*, 5, 301–314, 1972.

[13] Bradley, G.W., von Euler, C., Marttila, I. and Roos, B., 'A model of the central and reflex inhibition of inspiration in the cat', *Biol. Cybern.*, 19, 105–116, 1975.

[14] Saunders, K.B., Bali, H.N. and Carson, E.R., 'A breathing model of the respiratory system: the controlled system', *J. Theoret. Biol.*, 84, 135–161, 1980.

[15] Greer, W., Jordan, M.M. and Murray–Smith, D.J., 'A structural approach to respiratory control simulation', in Paul, J.P. et al. (ed.), *Computing in Medicine*, Macmillan Publishers, London, 339–346, 1982.

[16] Petersen, E.S. and Cunningham, D.J.C., 'Respiratory physiology', in Cramp, D.C. and Carson, E.R. (eds.) *The Respiratory System*, (Measurement in Medicine Series, Vol. 2), Croom Helm, London, 13–96, 1988.

[17] Grodins, F.S., 'Regulation of pulmonary ventilation', *The Physiologist*, 7, 319–333, 1964.

[18] Grodins, F.S., 'Exercise hyperpnea.   The ultra secret', in Hutas, I. and Debreczeni, L.A. (eds.), *Adv. Physiol. Sci., Vol. 10, Respiration*, Pergamon Press and Akademiai Kiado, Budapest, 243–251, 1981.

[19] Herczynski, R., 'The mathematical study of respiratory phenomena', in Cramp, D.G. and Carson, E.R., *The Respiratory System*, (Measurement in Medicine Series, Vol. 2), Croom Helm, London, 145–295, 1988.

[20] Murray–Smith, D.J. and Carson, E.R., 'The modelling process in respiratory medicine', in Cramp, D.G. and Carson, E.R., *The Respiratory System*, (Measurement in Medicine Series, Vol. 2), Croom Helm, London, 296–333, 1988.

[21] Murray–Smith, D.J. and Carson, E.R., 'Case studies of respiratory system models', in Cramp, D.G. and Carson, E.R., *The Respiratory System*, (Measurement in Medicine Series, Vol. 2), Croom Helm, London, 334–389, 1988.

[22] Milhorn, H.T., Jr., *The Application of Control Theory to Physiological Systems*, Saunders, Philadelphia, 1966.

[23] Milhorn, H.T., Jr. and Guyton, A.C., 'An analog computer analysis of Cheyne Stokes breathing', *J. Appl. Physiol.*, 20, 328–333, 1965.

[24] Longobardo, G.S., Cherniack, N.S. and Fishman, A.P., 'Cheyne–Stokes breathing produced by a model of the human respiratory system', *J. Appl. Physiol.*, 21, 1839–1846, 1966.

[25] Lange, R.L., Horgan, J.D., Botticelli, J.T., Tsagaris, T., Carlisle, R.P. and Kuida, H. 'Pulmonary to arterial circulatory transfer function: importance in respiratory control', *J. Appl. Physiol.*, 21, 1281–1284, 1966.

[26] Cherniack, N.S. and Longobardo, G.S., 'Cheyne–Stokes breathing – an instability in physiological control', *N. Engl. J. Med.*, 288, 952–957, 1973.

[27] Khoo, M.C.K., Knonauer, R.E., Strohl, K.P. and Slutsky, A.S., 'Factors inducing periodic breathing in humans: a general model', *J. Appl. Physiol.*, 53, 644–659, 1982.

[28] Nugent, S.T. and Finley, J.P., 'Periodic breathing in infants: a model study', *IEEE Trans.*, BME–34, 482–485, 1987.

[29] Pack, A.I. and Murray–Smith, D.J. (eds.), *Modelling the Respiratory System*, CRC Press, Boca Raton, Florida, (in press).

[30] Mills, R.J., Middleton, S., Moran, F., Murray–Smith, D.J. and Pack, A.I., 'Simulation in the teaching of concepts of respiratory gas exchange', *Int. J. Math. Educ. Sci. Technol.*, 5, 381–387, 1974.

[31] Murphy, T.W., *A two compartment model of the lung*, Rand Corp. Memo. RM–4833–NIH, 1966.

[32] Murphy, T.W., 'Modelling of lung gas exchange–mathematical models of the lung: the Bohr model, static and dynamic approaches', *Mathemat. Biosci.*, 5, 427–447, 1969.

[33] Emery, B., Moran, F. and Pack, A.I., 'A ventilation driven simulation of pulmonary gas exchange in man', *J. Physiol.*, 222, 38P, 1972.

[34] Pack, A.I., Emery, B., Moran, F. and Murray–Smith, D.J., 'Computer models of

gas exchange processes in pulmonary ventilation', in Wyke, B. (ed.), *Ventilatory and Phonatory Control Systems*, Oxford University Press, London, 227–247, 1974.

[35] Ferguson, D.R., Mills, R.J., Moran, F., Murray–Smith, D.J. and Pack, A.I., 'Estimation of the parameters of a lung model with clinical applications', in Eykhoff, P. (ed.), *Identification and System Parameter Estimation*, North Holland, Amsterdam, 1973.

[36] Bache, R.A. Gray, W.M. and Murray–Smith, D.J., 'Time–domain system identification applied to noninvasive estimation of cardiopulmonary quantities', *IEE Proc.*, 128, Part D, 56–64, 1981.

[37] Bache, R.A. and Murray–Smith, D.J., 'Structural and parameter identification of two lung gas–exchange models', in Vansteenkiste, G.C. and Young, P.C. (eds.), *Modelling and Data Analysis in Biotechnology and Medical Engineering*, North–Holland, Amsterdam, 175–188, 1983.

[38] Dickinson, C.J. *A Computer Model of Human Respiration*, MTP Press, Lancaster, England, 1977.

[39] Schwid, H.A., 'A flight simulator for general anethesia training', *Comput. Biomed. Res.*, 20, 64–75, 1987.

[40] Beneken, J.E.W. and Rideout, V.C., 'The use of multiple models in cardiovascular system studies: transport and perturbation methods', *IEEE Trans.*, BME–15, 281–289, 1968.

[41] Zwart, A., Smith, N.T., and Beneken, J.E.W., 'Multiple model approach to uptake and distribution of halothane: the use of an analog computer', *Comput. Biomed. Res.*, 5, 228–238, 1972.

[42] Riley, R.L. and Cournand, A., '"Ideal" alveolar air and the analysis of ventilation–perfusion relationships in the lungs', *J. Appl. Physiol.*, 1, 825–847, 1949.

[43] Bekey, G.A. and Yamashiro, S.M., 'Parameter estimation in mathematical models of biological systems', *Advances in Biomedical Engineering, Vol. 6*, Academic Press, New York, 1–43, 1976.

[44] Bekey, G.A. and Beneken, J.E.W., 'Identification of biological systems: a survey', *Automatica, 14*, 41–47, 1978.

[45] Murray–Smith, D.J., 'System identification and parameter estimation techniques in the modelling of physiological systems: a review', in Paul, J. et al. (eds.), *Computing in Medicine*, Macmillan, London, 333–338, 1982.

[46] Eykhoff, P. 'Biomedical identification: overview, problems and prospects', in Barker, H.A. and Young, P.C. (eds.), *Identification and System Parameter Estimation, 1985*, Pergamon Press, Oxford, 37–44, 1985.

[47] Rahn, H. and Fenn, W.O., *A Graphical Analysis of the Respiratory Gas Exchange*, American Physiological Society, Washington, 1955.

[48] Otis, A.B., 'Quantitative relationships in steady-state gas exchange', in Fenn, W.O. and Rahn, H., *Handbook of Physiology, Section 3: Respiration*, American Physiological Society, Washington, 1964.

[49] Murray-Smith, D.J. and Pack, A.I., 'Techniques of computer simulation applied to respiratory gas exchange', in Taylor, D.E.M. and Whamond, J.S. (eds.), *Non Invasive Clinical Measurement*, Pitman Medical, London, 186-202, 1977.

[50] Bekey, G.A. and Maloney, J.C., 'On-line estimation of cardiac output using respiratory models and measurements', *Fed. Proc.*, 29, 946 ABS.

[51] Maloney, J.C., *A New Non-Invasive Technique for Estimation of Cardiac Output from Respiratory Models and Measurements*, Ph.D. Dissertation, University of Southern California, 1971.

[52] Bellman, R. and Åström, K.J., 'On structural identifiability', *Math. Biosci.*, 7, 329-339, 1970.

[53] Beck, J.V. and Arnold, K.J., *Parameter Estimation in Engineering and Science*, Wiley, New York, 501, 1977.

[54] Goodwin, G.C. and Payne, R.L., 'Choice of sampling intervals', in Mehra, R.K. and Lainiotis, D.G., (eds.), *System Identification: Advances and Case Studies*, Academic Press, New York, 251-287, 1976.

[55] Federov, V.V., *Theory of Optimal Experiments*, Academic Press, New York, 1972.

[56] Brovko, O., Wiberg, D.M., Arena, L. and Belville, J.W., 'The extended Kalman filter as a pulmonary blood flow estimator', *Automatica*, 17, 213-220, 1981.

[57] Valcke, C.P., Jenkins, J.S., Ward, D.S. and Wiberg, D.M., 'Dynamic estimation of lung parameters', *Proc. IEEE/Eight Annual Conf. of Eng. in Med. and Biol. Society*, IEEE, Piscataway, New Jersey, 863-865, 1986.

[58] Jenkins, J.S., Valcke, C.P., Ward, D.S., and Wiberg, D.M., 'Modeling respired soluble gas uptake', *Proc. IEEE/Eighth Annual Conf. of Eng. in Med. and Biol. Society*, IEEE, Piscataway, New Jersey, 881-884, 1986.

[59] Kim, T.S., Rahn, H. and Farhi, L.E., 'Estimation of true venous arterial P_{CO2} by gas analysis of a single breath', *J. Appl. Physiol.*, 21, 1338-1344, 1966.

[60] Srinivasan, R., 'An analysis of estimation of pulmonary blood flow by the single breath method', *J. Appl. Physiol.*, 61, 198-209, 1986.

[61] Gronlund, J., 'Errors due to data reduction in single breath method for measurement of pulmonary blood flow', *J. Appl. Physiol.*, 52, 104-108, 1982.

[62] Sackner, M.A., Greeneltch, D., Heiman, M.S., Epstein, S. and Atkins, N., 'Diffusing capacity, membrane diffusing capacity, capillary blood volume, pulmonary tissue volume and cardiac output measured by a rebreathing technique', *Amer. Rev. Respir. Disease*, 111, 157-165, 1975.

[63] Weisiger, K.H. and Swanson, G.D., 'Importance of oscillations in alveolar gas

concentrations in the analysis of rebreathing data', *J. Appl. Physiol.*, <u>64</u>, 1104–1113, 1986.

[64] Prisk, G.K. and McKinnon, A.E., 'A modeling approach to the estimation of CO diffusing capacity', *J. Appl. Physiol.*, <u>62</u>, 373–380, 1987.

[65] Prisk, G.K. and McKinnon, A.E., 'Estimation of amount of stationary pulmonary blood from carbon monoxide uptake measurements', *J. Appl. Physiol.*, <u>63</u>, 1303–1308, 1987.

[66] Kapitan, K.S. and Wagner, P.D., 'Information content of multiple gas elimination measurements', *J. Appl. Physiol.*, <u>63</u>, 861–868, 1987.

[67] Drazen, J.M., Loring, S.H. and Regan, R., 'Validation of an automated determination of pulmonary resistance by electrical subtraction', *J. Appl. Physiol.*, <u>40</u>, 110–113, 1976.

[68] Neergaard, R. and Wirz, K., 'Uber eine Methode zur Messung der Lungenelastizitat am Lebenden Imenschen, Insbersondere beim Emphysema', *Z. Klein Med.*, <u>105</u>, 35–51, 1927.

[69] Dubois, A.B., Botello, S.Y. and Comroe, J.H., 'A new method for measuring airway resistance in a man using a body plethysmograph', *J. Clin. Invest.*, <u>35</u>, 327–335, 1956.

[70] Nada, M.D., 'Lung parameters identification', in Linkens, D.A. (ed.), *Biological Systems, Modelling and Control*, Peter Peregrinus, Stevenage, 242–274, 1979.

[71] Nada, M.D. and Linkens, D.A., 'An adaptive technique for estimating the parameters of a non–linear mathematical lung model', *Med. Biol. Eng. Comput.*, <u>15</u>, 149–54, 1977.

[72] Linkens, D.A. and Rimmer, S.J., 'On–line identification for mechanical parameters of the lung', *Trans. Inst. Meas. Contr.*, <u>4</u>, 177–185, 1982.

[73] Grimby, G., Takishima, T., Graham, W., Macklem, P. and Mead, J., 'Frequency dependence of flow resistance in patients with obstructive lung disease', *J. Clin. Invest.*, <u>47</u>, 1455–1465, 1968.

[74] Landser, F.J., Nagels, J., Demedts, M., Billiet, L. and Van de Woestyne, K.P., 'A new method to determine frequency characteristics of the respiratory system', *J. Appl. Physiol.*, <u>41</u>, 101–106, 1976.

[75] Michaelson, E.D., Grassman, E.D., and Peters, W.R., 'Pulmonary mechanics by spectral analysis of forced random noise', *J. Clin. Invest.*, <u>56</u>, 1210–1230, 1975.

[76] Baker, A.B. and Hahn, C.E.W., 'An analogue study of controlled ventilation', *Resp. Physiol.*, <u>22</u>, 227–239, 1974.

[77] Epstein, M.A.F. and Epstein, R.A., 'Airway flow patterns during mechanical ventilation of infants: a mathematical model', *IEEE Trans.*, <u>BME–26</u>, 299–306, 1979.

[78] Tatnall, M.L., 'The development of a system for the control of inhalational

anaesthesia', *Biomed. Meas. Infor. Contr.*, 1, 23–30, 1986.

[79] Lambertson, C.J. and Wendel, H., 'An alveolar pCO_2 control system: its use to magnify respiratory depresssion by meperidine', *J. Appl. Physiol.*, 15, 43–48, 1960.

[80] Bellville, J.W., Fleischli, G. and Attura, G., 'Servo control of inhaled carbon dioxide', *J. Appl. Physiol.*, 24, 414–415, 1968.

[81] Swanson, G.D., Carpenter, T.M., Jr., Snider, D.E. and Bellville, J.W., 'An on–line hybrid computing system for dynamic respiratory studies', *Comput. Biomed. Res.*, 4, 205–215, 1971.

[82] Swanson, G.D. and Bellville, J.W., 'Hypoxic–hypercapnic interaction in human respiratory control', *J. Appl. Physiol.*, 36, 480–487, 1974.

[83] Chambille, B., Guenard, H., Loncle, M. and Bar；eton, D., 'Alveostat, an alveolar P_{ACO2} and P_{AO2} control system', *J. Appl. Physiol.*, 39, 837–842, 1975.

[84] Murray–Smith, D.J., 'System identification and computer simulation – an integrated approach to the modelling of complex systems', in Vansteenkiste, G.C. et al. (eds.), *Proceedings of the Second European Simulation Congress, Sept. 9–12, 1986,* Society for Computer Simulation, San Diego, California, 3–9, 1986.

COMPUTER SIMULATION IN CANCER RESEARCH

Werner Düchting
Fachbereich Elektrotechnik, Universität Siegen,
Hölderlinstr. 3, D-5900 Siegen

1. The cancer problem

Already in old Egypt tumor diseases were well known. Since that time we have learned that cancer (neoplasm, tumor) is a disorder of cells and that its development is a multistep process. At least three different stages have been defined: Initiation, promotion and progression.

Characteristic features of malignant tumors are (1)
- uncontrolled proliferation
- invasion in adjacent normal tissue
- metastases induced to lymph nodes or other tissues via lymphatic channels or blood vessels
- ability to evade immune surveillance.

The classification of malignant tumors is based on
- the anatomical side of tumors
- histological classification of cells
- histological grade of malignancy
- size and degree of metastatic spread.

The majority of deaths by cancer are not caused by the primary tumor but by the growth of metastases resistant to therapy. This observation leads to the statement: The main problem in cancer treatment (by surgery, radiation-, chemo- or immunotherapy) is the prevention or the destruction of metastases (2). But in the background the central question which should be answered is: Which is the initiating event that is ultimately responsible for a stepwise transformation of a normal cell into a tumor cell?

In this century several hypotheses of the development of cancer have been the subject of intense investigation:
- virus theory (1910 by Rous)
- mutation theory (1914 by Boveri)
- metabolism theory (1926 by Warburg).

Today a unifying approach seems to become possible at the level of molecular biology. Recent investigations of the genetic alterations that cause normal cells to become malignant ones have focused on oncogenes. In this rapidly moving research area studies have revealed that dominant cellular genes called "proto-oncogenes" are activated by
- tumor viruses
- gene amplification
- gene translocation and
- genetic mutation.

However, the number of genetic alterations needed for neoplastic transformation remains to be elucidated. The state of the art may be described as follows: Experimental research work for diagnostic and therapeutic strategies is going on with the aim to identify human tumor-associated antigenes and to develop specific immunologic reagents directly against these targets. Thus, several monoclonal antibodies to detect and defeat carcinomas have now been generated. In spite of this progress one cannot yet identify genes of specific relevance to the multistep process of carcinogenesis. Therefore, the main question how genes and implicitly the growth of normal and malignant cells are regulated still remains open.

2. Biological observations of cell growth

In this section we try to summarize our current knowledge about the structure of cells and tissues and the mechanisms which control their growth.

2.1 Growth of normal cells

Each tissue has its own specific cells which maintain the structure and function of the individual tissue. There are different kinds of specific cell systems in existence
- permanent resting cell systems (brain, nerve, muscle cells)
- latent resting cell systems (liver cells)
- cell renewal systems (blood, epidermis, skin cells).

Most of the tissues in the body contain some cells that can renew themselves. A subset of the cell population in any tissue can differentiate the functional cells of that tissue. The normal process of cell differentiation leads to a fully differentiated cell that cannot, under normal circumstances, divide again (neurons, liver cells, kidney cells). If, however, a tissue is injured, the surviving cells with the capacity of divide begin to grow and to replace the damaged cells. When the replacement has been completed the repair process stops. It is assumed that there are stimulating and inhibiting factors which are normally in balance until a dividing stimulus is required. The division of a cell into two new ones involves four stages:

$$G1 \longrightarrow S \longrightarrow G2 \longrightarrow M.$$

- G1 is a gap after stimulation;   - G2 is as second gap period;
- S is the phase of DNA replication;   - M is the stage of mitosis.

At particular stages of the cell cycle the cells may be blocked, e.g. by drugs or agents, or they may move out of the cell cycle into a resting phase known as GO (3).

2.2 Growth of tumor cells

In contrast to the normal cell the tumor cell ist theoretically able to divide indefinitely. Furthermore, the tumor cell is a cell that has not achieved a fully differentiated state. It is blocked at some earlier stage of maturation. One can notice a different morphology, larger nucleus and an abnormal number of chromosomes in the cancer cell. Its feature is abnormal growth that means that a cell transformed by chemical, physical or biologic agents is no longer responsive to normal growth controlling mechanisms. It is assumed that oncogenes influence the transition from the GO- to the G1- and from the G1- to the S-phase of the cell cycle.

At the beginning of tumor growth the tumor is avascular, lacking its own network of blood vessels to supply oxygen and nutrients. This results in slow growth of the tumor. The critical event that converts a dormant tumor into a more rapidly growing neoplasm is associated with the vascularization of the tumor (4). The growth of new capillaries into the tumor is stimulated by a substance called tumor angiogenesis factor (TAF) produced by the tumor itself. The vascularized tumor begins to grow more rapidly. After local invasion of adjacent host tissue barriers, the tumor cells must invade the vascular wall in order to disseminate. They are carried to local lymph nodes or to distant organs where they produce metastases.

The cells in a single tumor are not all identical (homogeneous) but there is a range of cells with many differences in structure, morphology and growth rate. Therefore it is very difficult to study heterogeneous tumor growth in patients (in vivo). For studying the process of carcinogenesis tumors are induced to animals by applying carcinogenic agents or by tumor transplantation. Another way of observing growing tumor cells isolated from their normal environment is the cell culture technique (in vitro). Cell cultures cannot only be used to study the division of tumor cells, but also to determine the cytotoxic effect of chemotherapeutic drugs.

3. Possibilities of cancer treatment

We should mention ahead that all of the methods discussed here cannot cure patients with micrometastases.
- With some tumor diseases it is possible to remove the tumor by surgery.
- In other cases tumors are treated by radiation therapy. The mechanism how radiation kills cancer cells selectively is not yet fully understood. But all dividing (proliferating) tumor cells are particularly sensitive to radiation. The disadvantage of this method is that all proliferating normal cells will be damaged as well.
- Some of the anticancer drugs with toxic features act directly on tumor cells whereas others must be activated by metabolic processes. And many chemotherapeutic agents affect only a very particular phase of the cell cycle that means they act phase-specifically. The drugs have to be given on a definite schedule limited by toxicity. A very severe disadvantage of chemotherapeutic treatment is that tumor cells become resistant. Drug resistance belongs to the unsolved problems of chemotherapy.

- To treat cancer more efficiently combined approaches using surgery radiation- and chemotherapy are often applied to patients.
- In immunotherapy the immune system is stimulated by the administration of immunostimulating agents to respond to the tumor. On the other hand many experiments have been carried out to produce and use monoclonal antibodies of a desired specifity that will bind selectively to tumor for direct or indirect destruction of malignant cells. A disadvantage is that a monoclonal antibody against a tumor-associated antigen can also bind to some normal cells. But a positive argument should be pointed out: The application of immunotherapy extremely reduces the side effects and spares normal tissues.

4. Statement of the modeling problem

During the last years there has been a large progress in experiments gaining hard data about control processes in biology like temperature, blood pressure, cardiovascular, metabolism and cell growth control. On the other hand there is a wide arsenal of methods of systems analysis, control and automata theory and of computer science. Thus, it is near at hand to apply these theoretical methods to biological control systems.

In our case we focus our attention largely on normal and malignant cell growth systems. So far contributions to this topic mostly derive from the fields of cell biology, molecular biology, biochemistry, morphology and immunology. Therefore we try to open a new avenue by applying computer modeling and simulation to unsolved control problems in cancer research. The aim of our work is
- to construct models describing complex biological multi-loop cell growth control circuits (Fig. 1)
- to simulate their response to different perturbations
- to interpret structural unstable control loops as cancer
- to consider and to predict tumor growth in time and space (2-D, 3-D)
- to simulate different kinds of cancer treatment (surgery, radiation- and chemotherapy) in order to optimize the treatment strategies and schedules prior to clinical therapy.

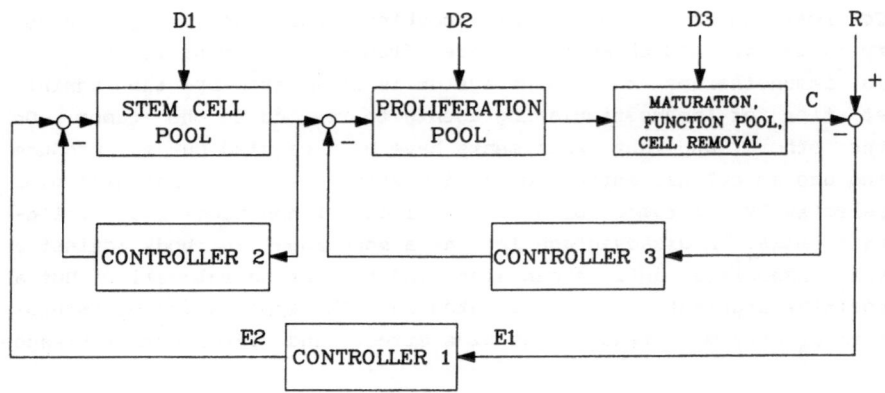

R: Required tissue oxygen (desired number of erythrocytes)

C: Number of red blood cells (erythrocytes)

E2: Production of the erythropoietin hormone

D1,D2,D3: Disturbances

Fig. 1: Multi-loop control circuit of erythropoiesis

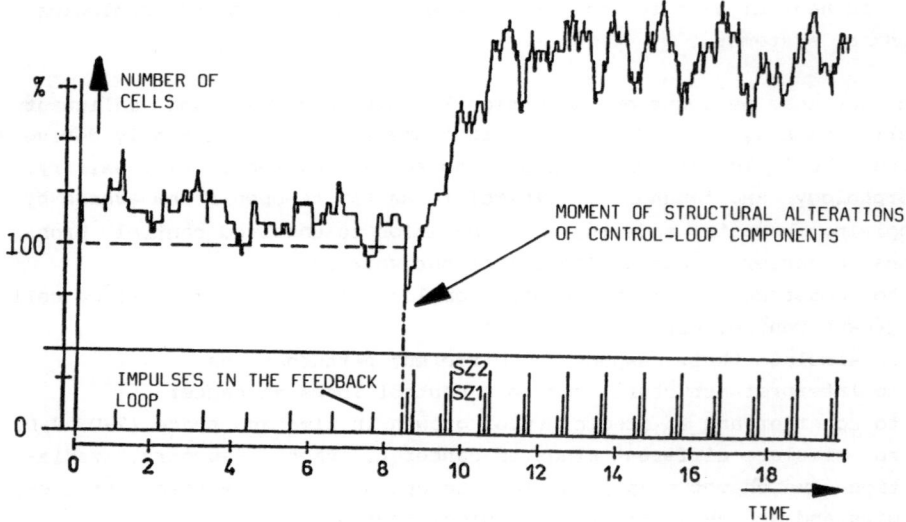

Fig. 2: Number of cells as a function of time indicating a cancerogeneous behaviour (simulation result)

5. Overview of previous work

In recent years a large body of dominantly mathematical oriented work applying mathematics to different areas of biological problems has been published (5-10). Only very few contributions considered biomedical questions from the viewpoint of control theory (11,12). Our own approach developing closed-loop control circuits for tumor growth started in 1968 (13). At that time the subject of consideration was focused on stability conditions. Step by step the dynamic behaviour of cell renewal control loops (Fig. 1) was investigated. Blockoriented simulation languages have been used for simulating the macromodels. As a result the number of cells as a function of time has been plotted (Fig. 2).

Then oncologists advised us to consider not only the time but also the spatial behaviour of tumor growth. In a first approach we developed models which described the 2-D behaviour of a normal cell inoculated into a nutrient medium (in a Petri dish). Next we extended this approach and tried to simulate tumor growth in the tissue of a tobacco leaf (14). After getting nice results (Fig. 3) we improved these models by introducing distinguished cell cycle phases (G1, S, G2, M, G0, N). Thus, we were able to simulate the 3-D growth of a single dividing tumor cell (15,16) inoculated into the center of the cell space at the beginning of the simulation run (Fig. 4).

As we know from section 2, the cytotoxic effect of chemotherapeutic drugs is tested in cell cultures. This is a very good in-vitro system which can be simulated by a computer model. Figure 5 shows the result of a test run in which all proliferating tumor cells (i.e. the outside rim) of the tumor spheroid have been killed by a cytotoxic drug at T=201 units of time.

After successful simulations of different in-vitro treatment (16) the attempt was made to substitute the nutrient medium by blood vessels that means to make a transition from in-vitro to in-vivo modeling and simulation. In the first approach fictious static capillaries have been introduced (17). Very soon it was clear that a more realistic structure of capillaries was desirable for simulating in-vivo growth of tumors. Therefore the next chapters of this contribution are relating to this difficult task.

123

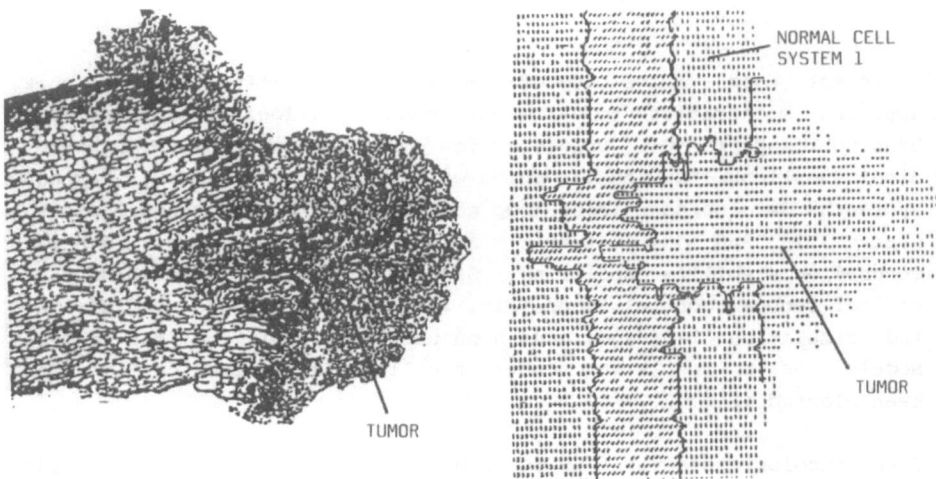

Fig. 3: Simulation of tumor formation in the tissue of a tobacco leaf

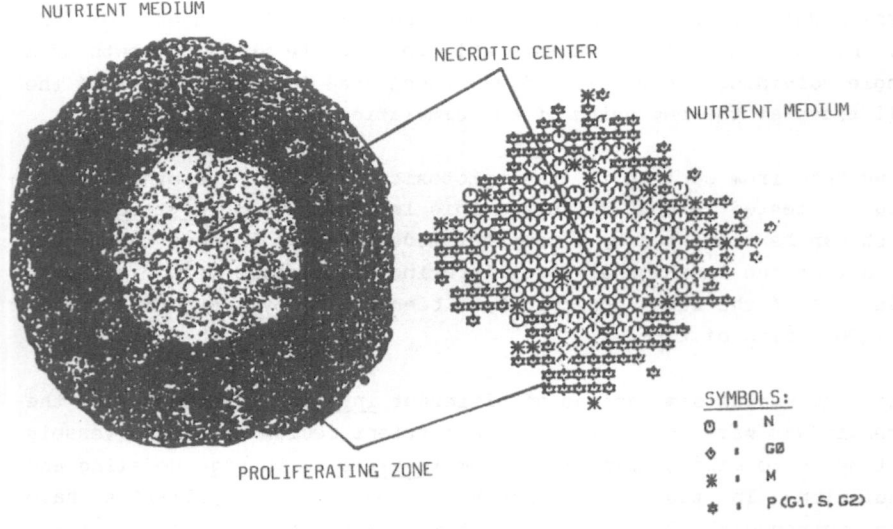

NUTRIENT MEDIUM

NECROTIC CENTER

NUTRIENT MEDIUM

PROLIFERATING ZONE

SYMBOLS:

◐	:	N
◈	:	GØ
✳	:	M
✴	:	P (G1, S, G2)

Fig. 4: Simulation of the formation of a tumor spheroid. The initial configuration consisted of a single tumor cell placed in the center of the nutrient medium

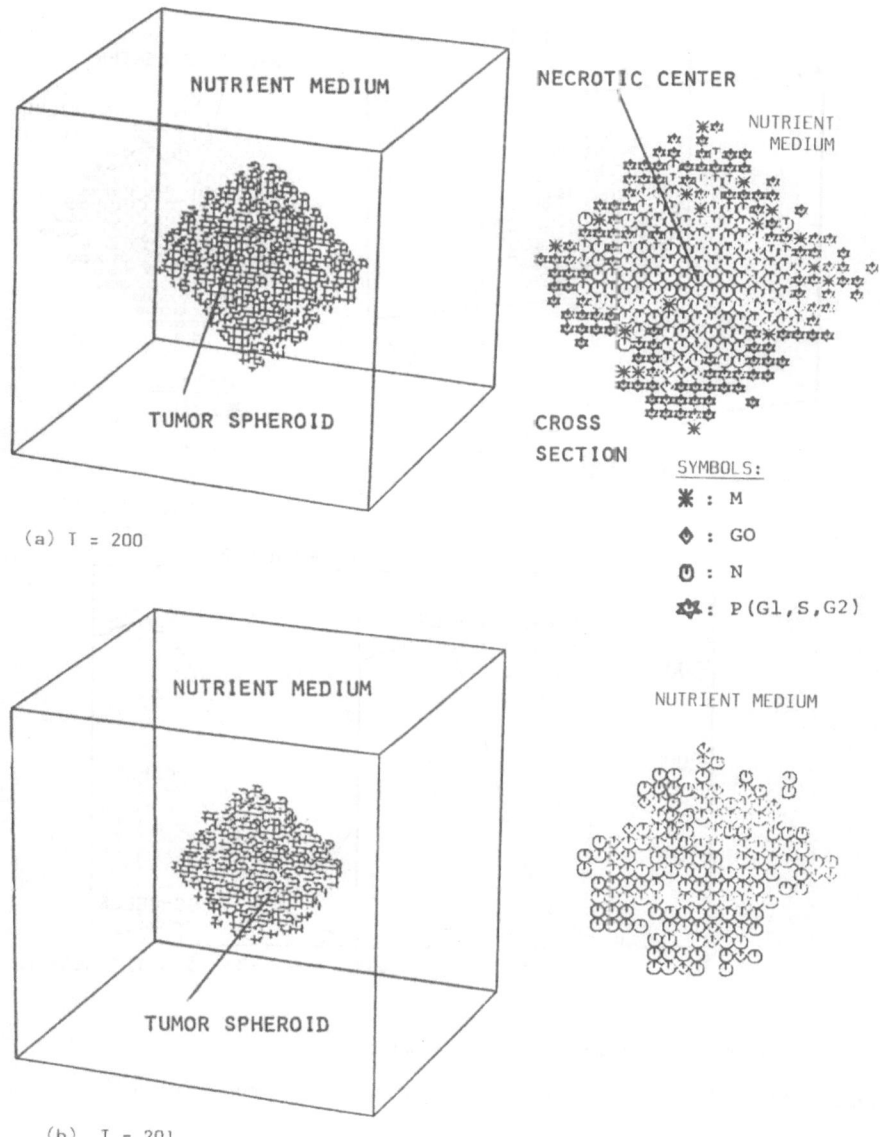

NUTRIENT MEDIUM

NECROTIC CENTER

NUTRIENT
MEDIUM

TUMOR SPHEROID

CROSS
SECTION

(a) T = 200

SYMBOLS:

✳ : M

◈ : GO

◍ : N

✿ : P(G1,S,G2)

NUTRIENT MEDIUM

NUTRIENT MEDIUM

TUMOR SPHEROID

(b) T = 201

125

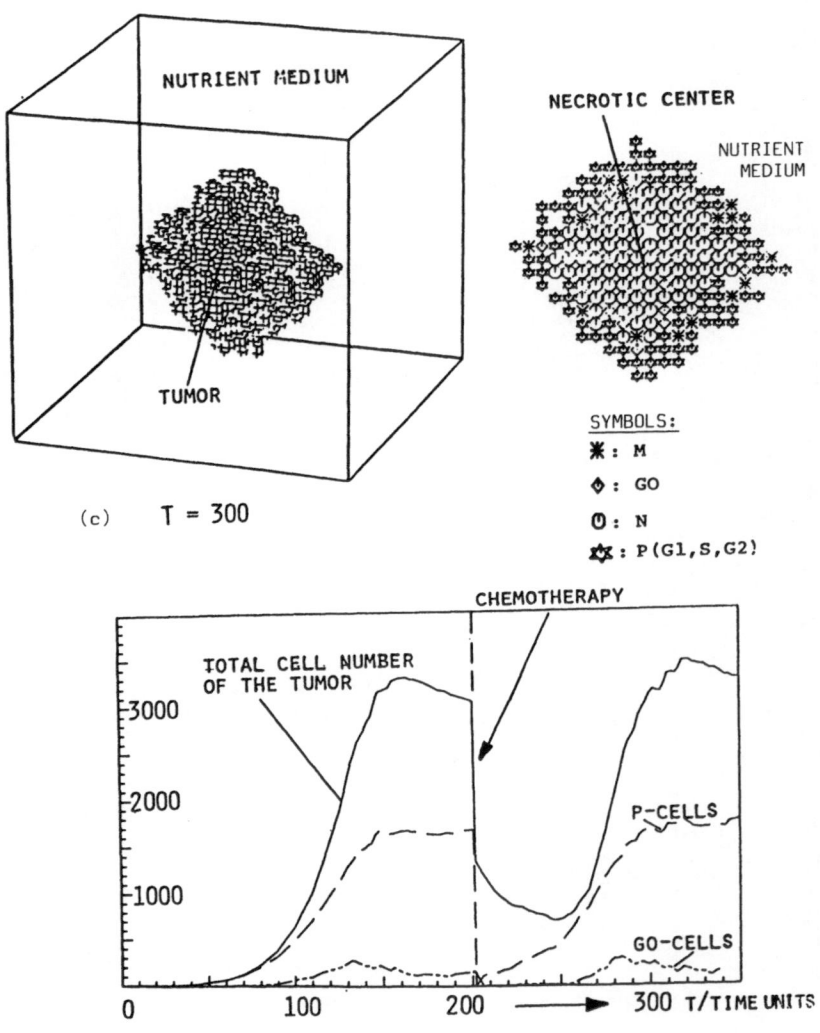

(c)　　T = 300

SYMBOLS:
※ : M
◈ : GO
Ⓞ : N
✿ : P(G1,S,G2)

(d) Number of tumor cells as a function of time

Fig. 5: Simulation of a chemotherapeutic treatment of a tumor spheroid
(in vitro)

126

6. Model design

6.1 Fundamental components

If you want to construct a model of high order as e.g. a complex
biological process, for instance cell growth, it is necessary to
design a modular concept. In this case it means to build modular
structured subsystems.

(i) At first you need control models which describe the cell division
of normal and tumor cells at a cellular level (Fig. 6).

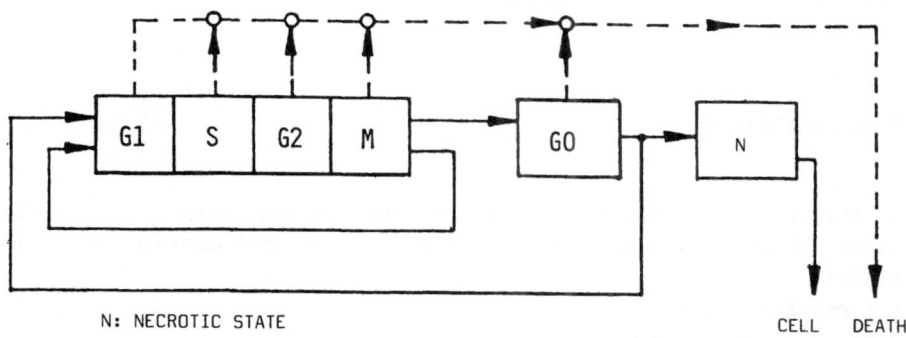

N: NECROTIC STATE   CELL   DEATH

Fig. 6: Simplified cytokinetic model of a tumor cell

(ii) Cell-production and interaction rules are required describing
the cell-to-cell communication.

One of several rules may be: The division of a tumor cell is possible
only if the distance between a dividing tumor cell and the blood
vessels is less than 100 µm equivalent to about the thickness of 3
cell layers. All tumor cells residing at a distance larger than 100
µm from the capillaries after the next division step will enter the
resting phase G0.

(iii) Cell movement is described by transport equations (diffusion-,
Poisson-equation). We have to introduce gradients for pressure and
metabolism for example

$$\xi_x = \frac{\partial p_x}{\partial x} \tag{1}$$

(p: pressure; ξ: pressure gradient)

or

$$v_x = f_x(k \frac{\partial c}{\partial x}) + s_x \tag{2}$$

    (v: speed of a moving cell; c: concentration of a metabolic compound; s: need for oxygen, glucose).

(iv) To represent 2-D and 3-D simulation results computer-graphics software-packages are necessary.

6.2 Limitations

At this stage of modeling complex cell growth some simplyfing assumptions about the real biological behaviour of cell division must be stated:
- a constant volume of a cubic cell
- constant phase duration and constant cell loss
- only horizontal and vertical correspondence of neighbouring cells
- a limited tissue volume by computer facilities
- side effects, immunologic reactions, heterogenity, drug resistance and the formation of metastases are neglected.

7. Computer-implementation of the model

The large body of statements, rules and equations previously discussed in section 6.1 has been transformed into algorithms. In addition algorithms considering tumor treatment (surgery, radiation- and chemotherapy) have been developed and shifted in subprograms written in FORTRAN IV. Figure 7 indicates their modular organization.

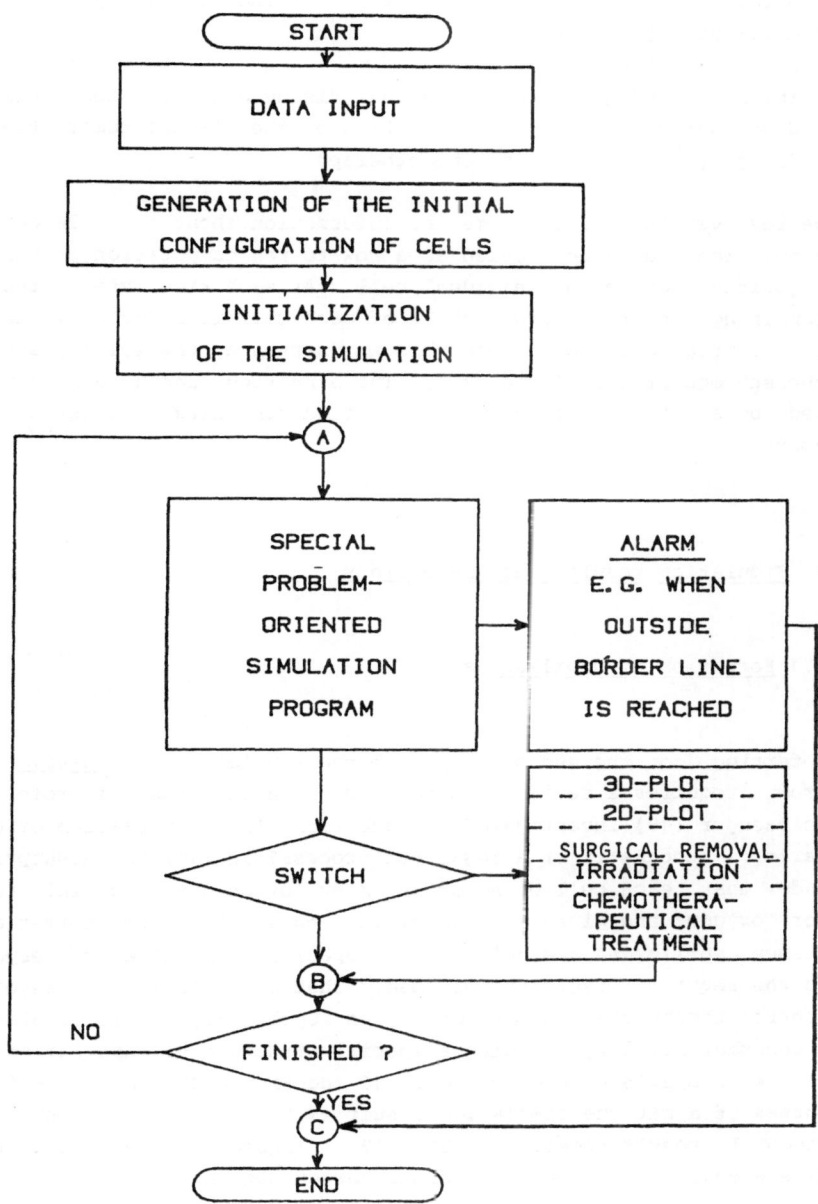

Fig. 7: Flow chart of the overall program run

129

To start the simulation program packages the following input data
have to be fed into the computer:
- notations about the character of a cell (normal, malignant)
- cell-cycle phase durations
- cell-loss rates
- initial configuration of normal tissue and of tumor cells
- distinguished data about the kind of the planned tumor treatment
  (surgery, radiation- and chemotherapy).

We reserved 64 bits to store the information about a single cell. The
simulation can be considered as a row-to-row computation of the cell
algorithm for each individual cell. At each time step, the time
remaining for the current phase of each cell is reduced by one unit.
The configuration gained in this way serves as the initial state of
the subsequent calculation step. The simultion runs have been performed on a VAX-730 machine. A simulation run takes at least about 12
hours.

8. Simulation results and discussions

8.1 Formation of capillaries

Referring to the end of chapter 5 the simulation of _in-vivo_ tumor
growth requires a realistic structure of capillaries. Therefore Vogelsaenger (18) investigated the question: Is the formation of capillaries a stochastic or a regulated process? In (18) the assumption is
made that each cell of an organ in evolution has a special request
for oxygen and glucose. Therefore, parallel to the formation of
tissue capillaries are built with a specific structure corresponding
to the required oxygen and glucose. That means from the viewpoint of
control theory the request for oxygen supply (Fig. 8) is regulated to
a constant level by building a special structure of capillaries (Fig.
9). A comparison between Figure 10 and Figure 11 shows that for the
cortex of a rat the simulation result is highly similar to the experimental result received by Bär (19). Figure 12 gives you a three-
dimensional impression of a normal cortex segment.

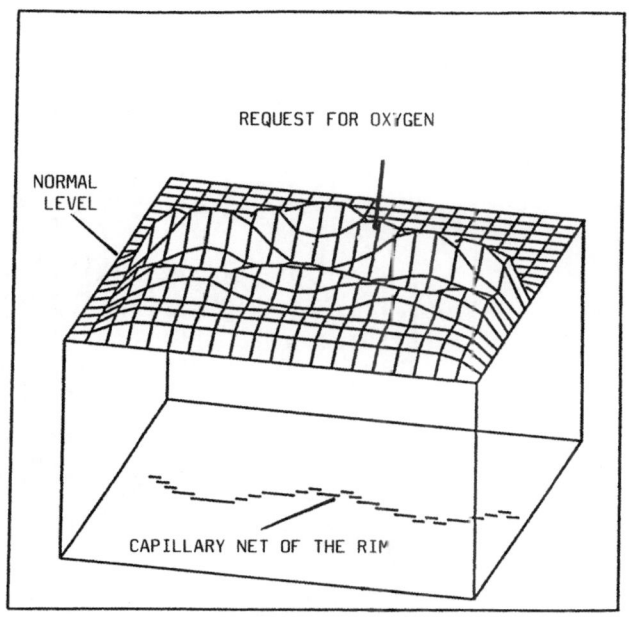

Fig. 8: Request for oxygen supply in the cortex

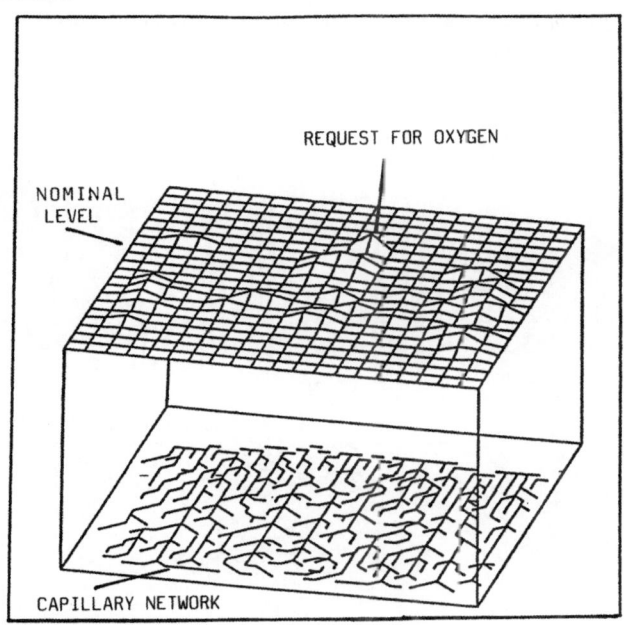

Fig. 9: Formation of capillaries in the cortex

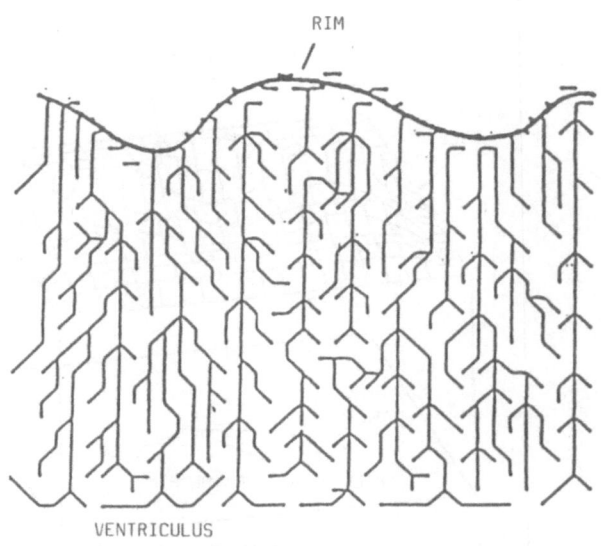

Fig. 10: Capillary network in the cortex (simulation result)

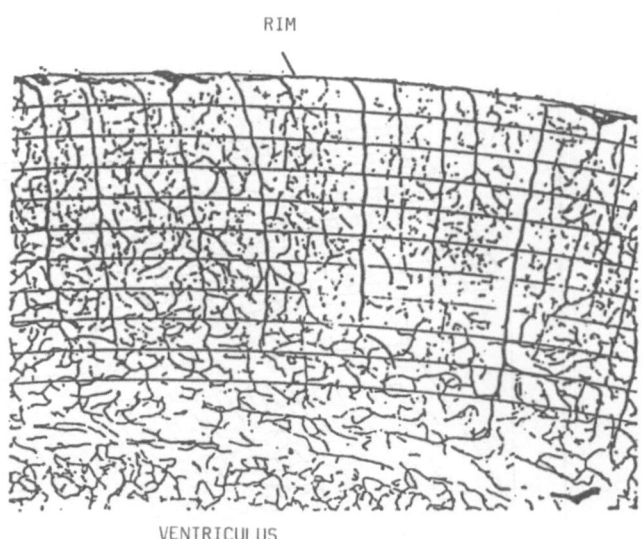

Fig. 11: Vascularization of the cortex (19)

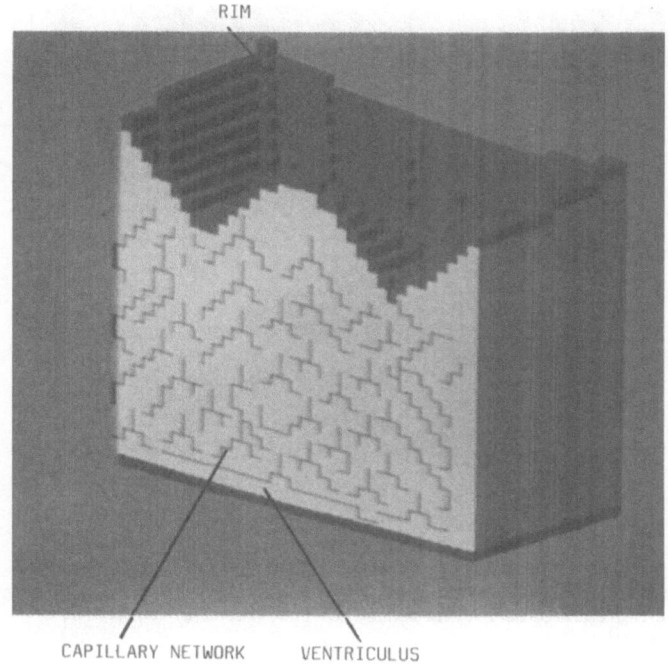

RIM

CAPILLARY NETWORK   VENTRICULUS

Fig. 12: **Three-dimensional representation of a cortex segment**

8.2 Spread of tumor cells in the cortex

Now the assumption is made that a single tumor cell is arbitrarily
placed in the tissue of the cortex at T=1 unit of time (Fig. 13). If
this tumor cell resides close to a capillary it will divide and move
in accordance with the cell production rules described in section
6.1. In a larger distance from the capillaries tumor cells will
transit into the nectrotic state as Carlsson (20) has observed (Fig.
14). This can be seen in the series of representations characterizing
the spread of the tumors cells (Fig. 15 and Fig. 16). The dark tumor
cells are symbolizing their necrotic state (21).

133

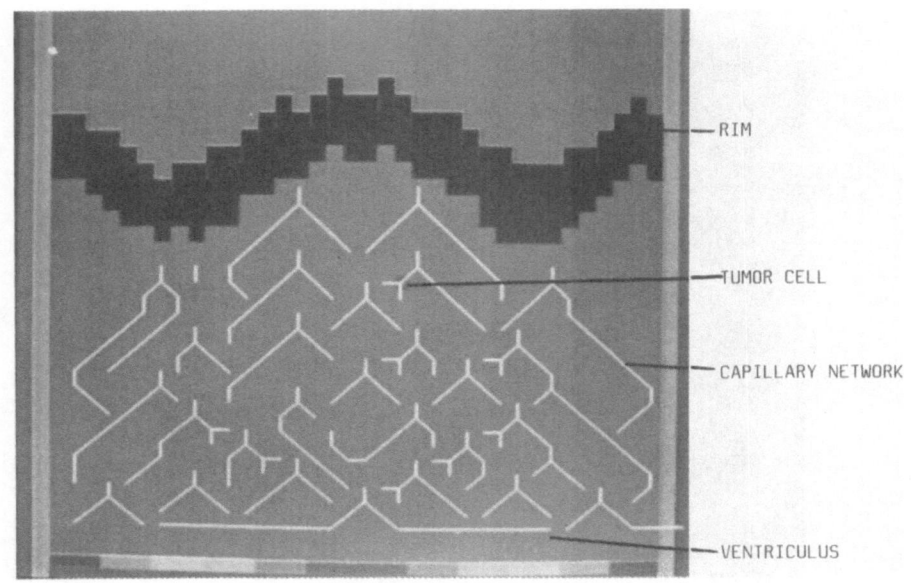

Fig. 13: A single tumor cell is introduced in the tissue of the cortex
(T = 1 unit of time)

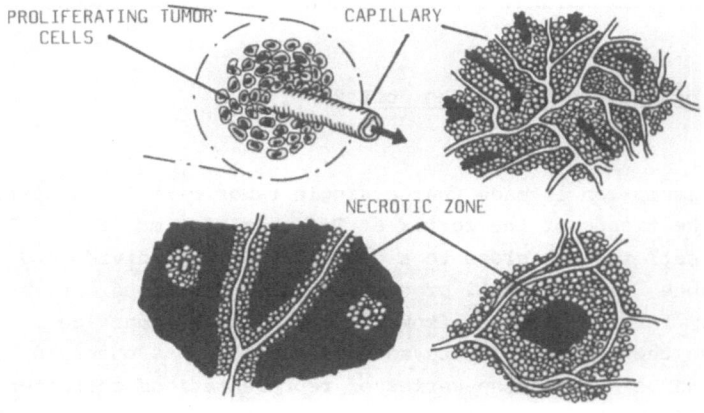

Fig. 14: Capillaries surrounded by proliferating tumor cells (20)

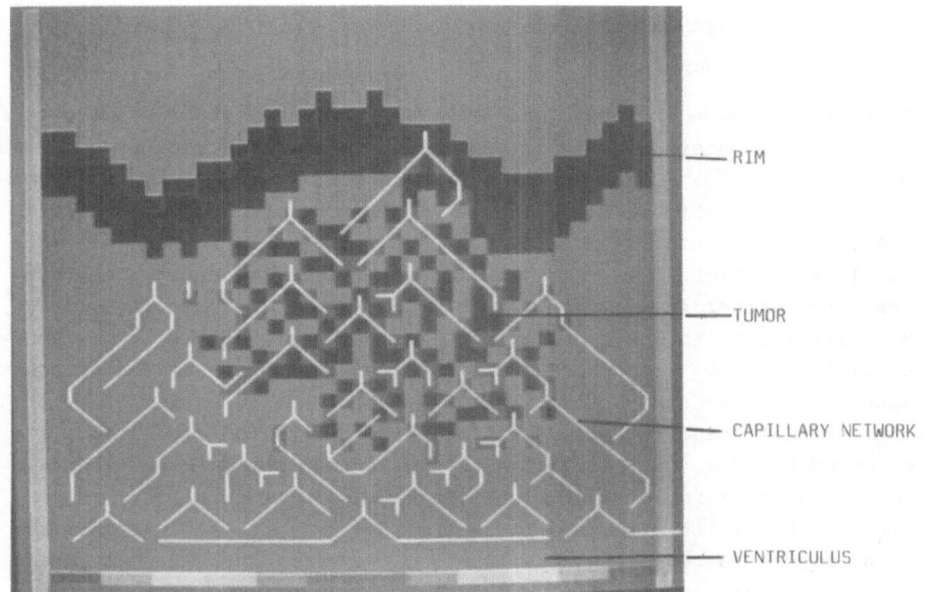

Fig. 15 : Spread of tumor cells in the cortex (T≠ 45 units of time)

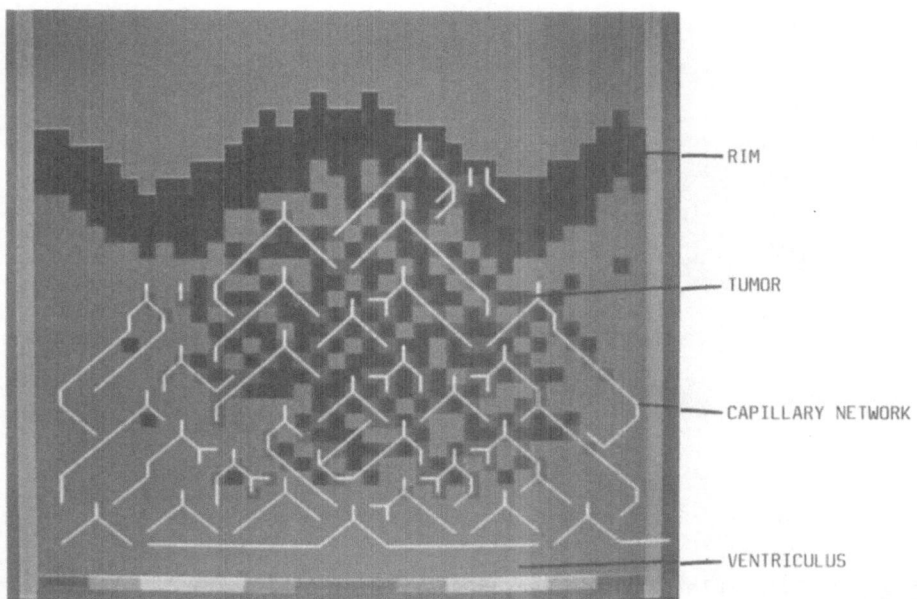

Fig. 16: Spread of tumor cells in the cortex (T=100 units of time)

8.3 Tumor angiogenesis effect

Folkman (4) made the observation that a spheroidal tumor stops growing at a diameter of a few millimeters because of lack of oxygen and
nutrients supply. Further tumor growth is possible only because tumor
cells produce a substance which is called tumor angiogenesis factor
(TAF). This factor stimulates nearby blood vessels to send out new
capillaries (Fig. 17) which grow towards the tumor, penetrate it and
lead to further rapid tumor growth. This tumor angiogenesis effect
was simulated in our computer model with a very simple additional
assumption: The rapid spread of tumor cells requires an additional
amount of oxygen and glucose. This request cannot be satisfied by the
conventional capillaries. Therefore the simulation algorithms are
extended to consider that if the oxygen request passes over a certain
value the formation of additional capillaries will be started according to the rules discussed in section 8.1. Figure 18 in comparison
with figure 16 represents the growth of new capillaries that means
the tumor angiogenesis effect. Recently great efforts have been made
to attack cancer by trying to find a protein which inhibits the
production of the tumor angiogenesis factor.

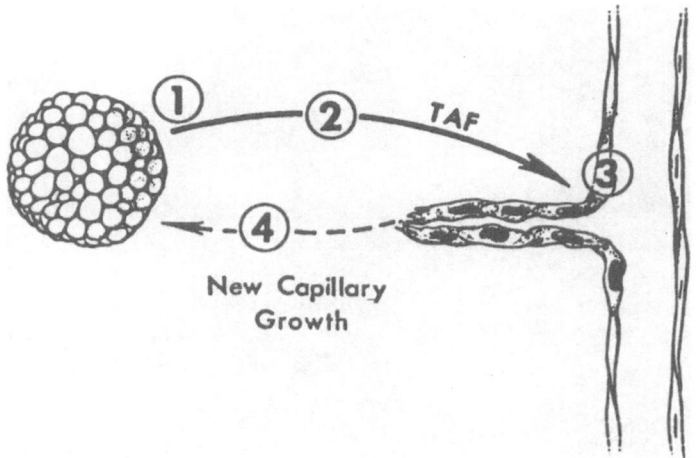

Fig. 17: Principle of tumor angiogenesis effect (4)

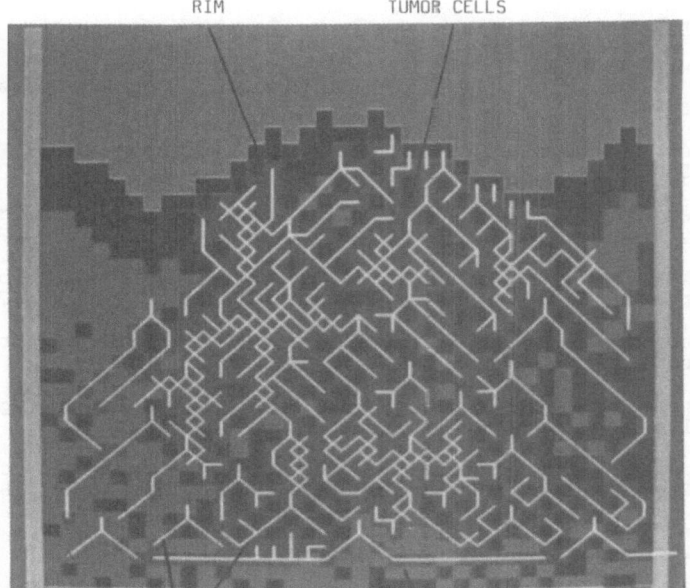

RIM   TUMOR CELLS

NEW CAPILLARIES   VERTRICULUS

Fig. 18: Tumor angiogenesis effect: Formation of new capillaries
(T=120 units of time)

9. Outlook

There are some promising avenues of future research work.

- The consideration of facts which had to be neglected so far (formation of metastases, immunologic reactions, drug resistance, heterogenity, side effects).
- The partial substitution of long and expensive biological test series by computer simulation.
- The generation of a more realistic initial configuration of a tumor
  by combining CT-pictures (Computer Tomography) with our predictive
  models describing tumor growth (22,23).
- The optimization of distinguished methods and schedules of cancer
  treatment.

10. References

(1) Franks, L.M. and Teich, N. (eds.), Introduction of the Cellular and Molecular Biology of Cancer, Oxford University Press, Oxford 1986.

(2) Weiss, L., Principles of Metastasis, Academic Press, Orlando 1985.

(3) Baserga, R., The Biology of Cell Reproduction, Harvard University Press, Cambridge/Mass. 1985.

(4) Folkman, J., How Is Blood Vessel Growth Regulated in Normal and Neoplastic Tissue?, Cancer Research, Vol. 46 (1986): 467-473.

(5) Jäger, W., Rost, H. and Tautu, P. (eds.), Biological Growth and Spread, Springer-Verlag, Berlin 1980.

(6) Wichmann, H.-E. and Loeffler, M., Mathematical Modeling of Cell Proliferation, CRC Press, Boca Raton / Florida 1985.

(7) Eisenfeld, J. and DeLisi, C. (eds.), Mathematics and Computers in Biomedical Applications, North-Holland, Amsterdam 1985.

(8) Segel, L., Modeling Dynamic Phenomena in Molecular and Cellular Biology, Cambridge University Press, Cambridge 1984.

(9) Wolfram, St., Cellular Automata as Models of Complexity, Nature, Vol. 311, No. 5985 (1984): 419-424.

(10) Meinhard, H., Models of Biological Pattern Formation, Academic Press, London 1982.

(11) Swan, G.W., Applications of Optimal Control Theory in Biomedicine, Marcel Dekker Inc., New York 1984.

(12) Möller, D., Popovic, D. and Thiele, G., Modeling, Simulation and Parameter-Estimation of the Human Cardiovascular System, Friedr. Vieweg & Sohn, Braunschweig/Wiesbaden 1983.

(13) Düchting, W., Krebs, ein instabiler Regelkreis, Versuch einer Systemanalyse, Kybernetik, 5. Band, 2. Heft (1968): 70-77.

(14) Düchting, W. and Dehl, G., Spatial Structure of Tumor Growth: A Simulation Study, IEEE Transactions on Systems, Man and Cybernetics SMC-10, No. 6(1980): 292-296.

(15) Mueller-Klieser, W., Multicellular-Spheroids, J. Cancer Res. Clin. Oncol. 113 (1987): 101-122.

(16) Düchting, W. and Vogelsaenger, Th., Three-Dimensional Pattern Generation applied to Spheroidal Tumor Growth in a Nutrient Medium, Int. J. Bio-Medical Computing 12(1981): 377-392.

(17) Düchting, W. and Vogelsaenger, Th., Aspects of Modelling and Simulating Tumor Growth and Treatment, J. Cancer Res. Clin. Oncol. 105(1983): 1-12.

(18) Vogelsaenger, Th., Modellbildung und Simulation von Regelungs-mechanismen wachsender Blutgefäßstrukturen in normalen Geweben und malignen Tumoren, Dissertation Siegen, 1986.

(19) Bär, Th., Patterns of vascularization in the developing cere-bral-cortex, CIBA Found. Symp. 100(1983): 20-36.

(20) Carlsson, J., Tumor models in vitro: A study of proliferation and growth in cellular spheroids, Acta Univ Upsal 466, 1978.

(21) Düchting, W., Simulation of 3-D Tumor Growth and Radiation Therapy in "Proceedings of the International Symposium Computer Assisted Radiology" edited by H.U. Lemke, M.L. Rhodes, C.C. Jaffee and R. Felix, Springer-Verlag, Berlin 1987: 335-339.

(22) Greinacher, C.F.C., Luetke, B. and Seufert, G., Digital Image Information Systems in Radiology, Siemens Forsch.- u. Entwickl.-Ber. Bd. 16, Nr. 1 (1987): 22-29.

(23) Höhne, K.H., Riemer, M. and Tiede, U., Viewing Operations for 3-D-Tomographic Gray Level Data in "Proceedings of the Internatio-nal Symposium Computer Assisted Radiology" edited by H.U. Lemke, M.L. Rhodes, C.C. Jaffee and R. Felix, Springer-Verlag, Berlin 1987: 599-609.

MATHEMATICAL SIMULATION OF THE HUMAN
THERMAL SYSTEM

Jürgen Werner
Ruhr-Universität, Institut für Physiologie
Abt. Biokybernetik, MA 4/59, D 4630 Bochum 1
Federal Republic of Germany

1. Introduction

Burton (7) was the first to calculate system properties of the human thermal system on the basis of a very general heat balance. With increasing physiological knowledge and increasing availability of more and more efficient analog and digital computers more sophisticated approaches have been proposed during the past 20 years.

The current efforts in this field may be characterized by the attempt to take into account one or more of the following aspects:
(1) geometry and inhomogeneity of the body
(2) heat transport via circulation
(3) regulatory concepts and strategies
(4) interaction with other regulatory systems.

Reviews on the historic development have been given by Hardy (18), Hwang and Konz (25), Wissler (63, 64), Werner (56, 58) and others.
For this reason only few contributions will be mentioned here without underestimating the meritorious publications of many other colleagues. Atkins and Wyndham (4, 66) first took into account differences of the thermal properties of different tissues by using a one-cylinder analog computer model with four layers. Wissler (59, 60, 61, 62) used six and more cylindrical elements and computed steady-state radial temperature profiles on digital computers. A further more precise characterisation of the system was achieved by introducing models with distributed parameters by Werner (51, 53), Kuznetz (31) and Wissler (64).

Convective heat transport by the circulatory system had mostly to be treated in a simplified way e.g. by Wissler (61, 62, 63) and Jiji and co-workers (28, 29).

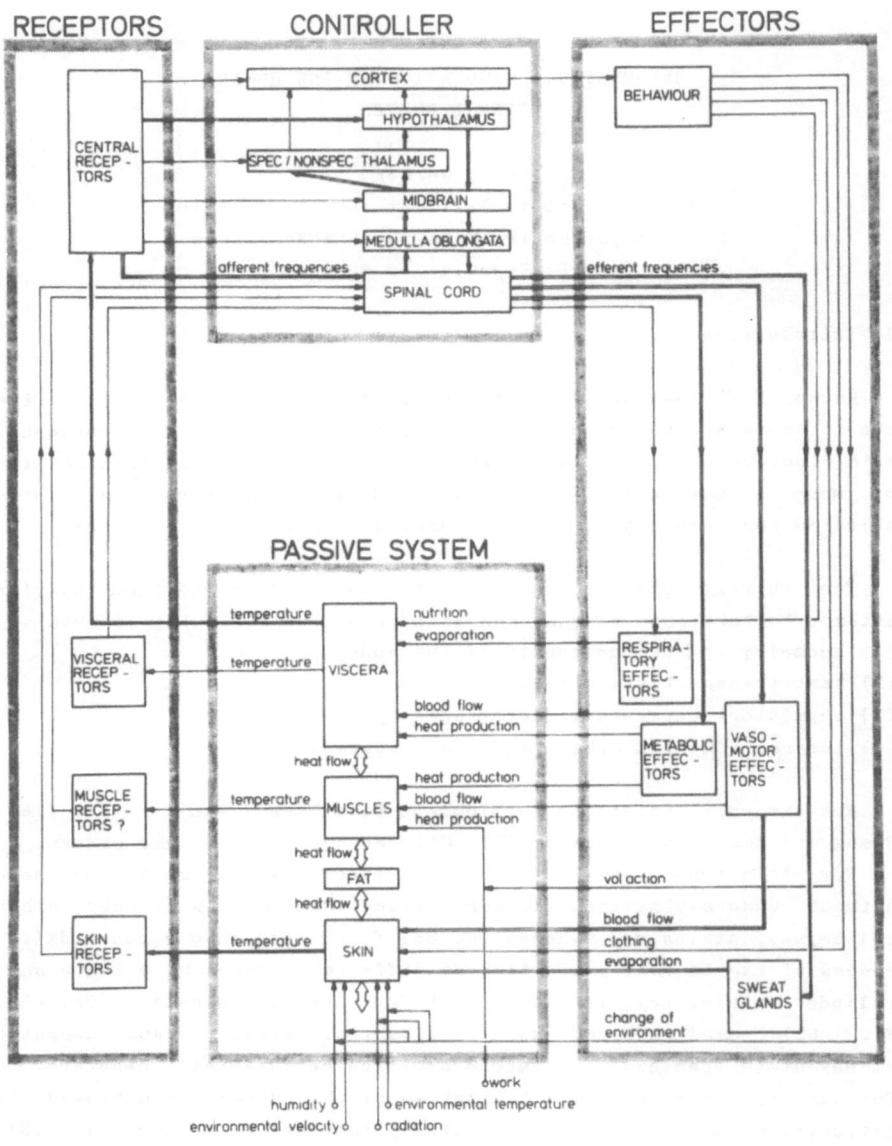

Fig. 1   The current view of the thermoregulatory system.

142

The early models analyzed the passive unregulated system, whereas Crosbie and co-workers (13) first took into account regulatory feedback mechanisms. A further decisive development of modeling the controlling elements was presented by Stolwijk and Hardy (44). This paper is still the basis of many more elaborate models (Atkins and Wyndham (4), Stolwijk (45), Kuznetz (30), Werner (50)). Recent models by Wissler (62) draw attention to the interaction with circulatory and respiratory regulation. Beside the heat balances for body and circulation, Wissler's recent models take into account for example mass balances for O_2, CO_2 and lactate.

Figure 1 gives a survey on the current view of human temperature regulation, from which some essential control characteristics should already proceed.

Temperature-sensitive elements are thought to be heterogeneously distributed throughout the body. The existence of warm and cold receptors in the skin is well known. However, there are still some doubts as to the presence of temperature-sensitive elements in skeletal muscles (27).

Both Rawson and Quick (38) and Riedel and co-workers (40) demonstrated the change of effector activity by thermal stimulation of the visceral muscles. According to the results of Jessen and co-workers (27), it is evident that apart from the well-known sensitive areas, skin and CNS, there are most likely other sensitive elements in body, which taken together should have the same importance as those in the CNS. However, Hellon and co-workers (21) have not so far succeeded in finding thermosensitive elements in the main blood vessels. As to the CNS itself, thermosensitivity can be demonstrated almost anywhere, the predominant areas being the spinal cord and the hypothalamus. Mentioning the large variety of important papers dealing with neuronal extrahypothalamic thermosensitivity would exceed the scope of this paper. Fortunately this is done by several reviews (see for example Hensel (23); Simon (42)). The current idea of the thermoafferent system and of the integrating centres is also shown in fig. 1: spinal cord, medulla oblongata, midbrain, specific and non-specific thalamus, hypothalamus and cortex.

The efferent part of the system includes autonomous and behavioural effector mechanisms. Thermoregulatory behaviour is affected by nutrition, voluntary muscle activities, clothing, change of environment and so on. The autonomous effectors are, above all, heat production via metabolism, skin blood flow via vasomotor activity and evaporation via sweat production and respiration (for details see for example by Cabanac (8) or Werner (56)).

Regarding the passive system, four compartments are taken into account in Fig. 1: the insulating shell consisting of skin and fat, and the core areas (muscles and viscera). Environmental factors are temperature, humidity, air velocity, and radiation; an important endogenous factor is heat production due to work load.

The current concept of the system has thus to take into account locally distributed measurement, processing and actuation.

2. Distributed Parameter Model: Radial Dependencies

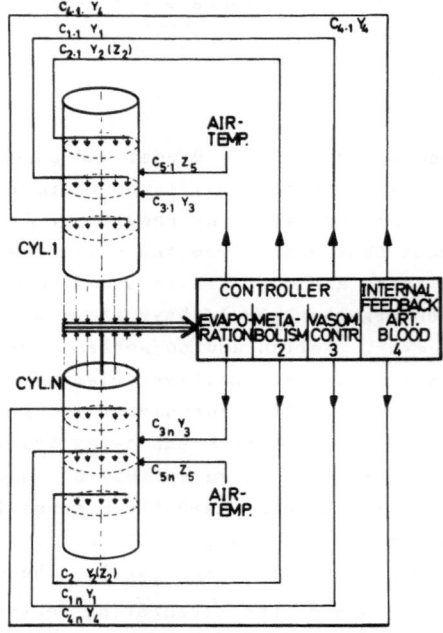

Fig. 2   Multi-element (circle)-model.

144

These types of models take into account the radial dependency of the essential control variables. The set of partial differential equations may be solved by numerical methods or after linearization by direct methods (Werner (50, 51)). Due to the difficult geometry of human body some simplifications are done. The body is devided into n cylindrical elements (Fig. 2), and the practical calculations are carried out by the example of n = 10: head, trunk, arms, hands, legs, feet.

We assume a one-loop circulatory system. A central pool of blood at temperature T_a delivers the arterial blood to capillaries and tissue (temperature T_i). Through the veins blood at temperature T_{vi} flows back to the central pool.

The instationary heat-flow is affected by three mechanisms within the passive system: metabolism, conduction through the tissue and convection through the blood. Assuming a purely radial conduction within the cylindrical elements, we get the following equation for each area of constant conductivity index λ_i (i = 1...n). (The differential and integral operators to be used are restricted to the intervals of continuously changing parameters):

Instationary heat flow = metabolism
+ conduction through tissue
+ heat-flow-input through blood
− heat-flow-output through blood

$$\rho_i c_i \frac{\partial T_i}{\partial t} = M_i + \lambda_i \left\{ \frac{\partial^2 T_i}{\partial r^2} + \frac{1}{r} \frac{\partial T_i}{\partial r} \right\} + $$

$$\beta_i Q_i \rho_a c_a T_a - \beta_i Q_i \rho_a c_a T_i \quad i = 1...n \tag{2.1}$$

with

t = time

r = radial coordinate

$T_i(r,t)$ = temperature of tissue plus capillaries in cylinder i

$M_i(r,t)$ = heat-production by metabolism per volumetric unit

$Q_i(r,t)$ = blood-flow per volumetric unit

$T_a(t)$ = temperature of central blood (lungs, heart, large, vessels)

$T_A(t)$ = air-temperature

$E_i(t)$ = heat-loss by evaporation

$T_{vi}(t)$ = temperature of venous blood

$R_R(t)$ = respiratory heat-loss
$\rho_i(r)$ = density of tissue plus capillaries
$c_i(r)$ = specific heat of tissue plus capillaries
$\lambda_i(r)$ = thermal conductivity index
β_i = counter-current-factor
ρ_a = density of central blood
c_a = specific heat of central blood
r_{si} = outer radius of cylinder i
h_i = heat-transfer-coefficient
A_i = surface
m_a = mass of central blood

Initial conditions: $T_i(r,o)=T_{io}(r)$ (2.2)
Boundary conditions: $r = r_{si}$ (skin):

$$\left\{ \lambda_i \frac{\partial T_i}{\partial r} = h_i(T_A - T_i) - \frac{E_i}{A_i} \right\}_{r\,=\,r_{si}} \qquad (2.3)$$

$$r = 0: \quad T_i \text{ definite} \qquad (2.4)$$

Temperature of central blood is considered independent of radius, but dependent on time. So we get another heat-balance-equation:

$$m_a c_a \frac{\partial T_a}{\partial t} = \sum_{i=1}^{n} \left\{ \beta_i \rho_a c_a Q_{ig} (T_{vi} - T_a) \right\} - R_R \qquad (2.5)$$

Q_{ig} is blood-flow, integrated over the cylindrical volume. From the heat-balance for the blood flowing back into the central pool one gets an equation for the temperature of the venous blood T_{vi}.
The passive uncontrolled system already possesses an inner feedback via circulation (cf. Fig. 2).
For formal reasons this feedback-loop is treated later together with the regulatory feedbacks. It is convenient to compute the linearisation for the product QT, so that a linear system is obtained for deviations ϑ_i from stationary temperatures. The equations are arranged, so that on the left we have all terms dependent on ϑ_i and on the right all terms with effector- and disturbance-variables. The right side of the differential equations is then the so-called source-function Φ_{Qi}, and the right side of the boundary-condition the boundary-function Φ_{Ri}.

We admit that parameters may change discontinuously at s - 1 isolated
points. Within the j = 1...s intervals we assume constant values e.g.
λ_{ji} (conductivity index), Q_{oij} (basal blood flow), etc. At the points
of discontinuity r_{ji} we get two additional conditions:

$$\vartheta_i(r_{ji}-o) = \vartheta_i(r_{ji}+o) \tag{2.6}$$

$$\lambda_{ji}\left(\frac{\partial\vartheta_i}{\partial r}\right)_{r_{ji}-o} = \lambda_{j+1,i}\left(\frac{\partial\vartheta_i}{\partial r}\right)_{r_{ji}+o} \tag{2.7}$$

with j = 1...(s - 1).

The equations are solved by way of two integral-transformations. The
first of them runs as follows, using the coordinate r in the normed
form (0...1):

$$\vartheta_i^*(\gamma_i,t) = \int_0^1 \vartheta_i(r,t)K_i(r,\gamma_i)dr \tag{2.8}$$

with the still unknown kernel $K_i(r,\gamma)$. Φ_{Qi} and Φ_{Ri} are transformed in
an analogous manner. The transformation of the whole differential
equation yields an eigen-value-problem with eigen-values k_i, which
permits the evaluation of the kernel K_i from a Bessel-equation of
order zero. The solution of the eigen-value-problem provides the
complete orthonormal system of eigen-functions M_{im}, which are closely
related to the kernel-functions. For each interval of continuous
parameters we get cylinder-functions with argument $\gamma_{im}r$, γ_{im} being
computed directly from the eigen-values k_{im}.

$$\text{Eigen-functions: } M_{im} = \sqrt{\alpha_{im}}\ C_o(\gamma_{im}r) \tag{2.9}$$

α_{im} is the normalizing factor to achieve orthonormality. The eigen-
functions obey the homogeneous boundary and transition-conditions
for ϑ_i.
Taking this into account the partial differential equation is converted through transformation into an ordinary differential equation:

$$\frac{\partial}{\partial t}\vartheta_{im}^* + k_{im}\ \vartheta_{im}^* = \Phi_{Qim}^* + \Phi_{Rim}^* \tag{2.10}$$

Equ. (2.10) is submitted to another well-known integral-transforma-
tion, the Laplace-transformation (variable p). Just for simplicity we
do not choose another symbol for the transformed variables.
For disappearing initial-values we obtain:

$$\vartheta^*_{im} = \frac{1}{k_{im} + p} \{\Phi^*_{Qim} + \Phi^*_{Rim}\} \tag{2.11}$$

Equ. (2.4) is equally transformed after linearisation and yields an equation for the deviation of central blood temperature.
We expand the system-output $\vartheta_i(r,p)$ in a Fourier-series of the eigen-functions.

$$\vartheta_i(r,p) = \sum_{\mu=1}^{\infty} \vartheta^*_{im} \sqrt{\alpha_{im}} C_0(\gamma_{im}r) \tag{2.12}$$

After substituting ϑ^*_{im} by equ. (2.11) and equally Φ^*_{Qim} and Φ^*_{Rim} we get, using the integration variable η:

$$\vartheta_i(r,p) = \sum_{\mu=1}^{\infty} \frac{\alpha_{im}}{k_{im}+p} \cdot \tag{2.13}$$

$$\int_0^1 \Phi_{Qi}(\eta,p)\eta C_0(\gamma_{im}\eta) C_0(\gamma_{im}r)\varsigma_i c_i \, d\eta + \Phi_{Ri} C_0(\gamma_{sim})\varsigma_{si}c_{si}C_0(\gamma_{im}r)$$

The theory of distributed-parameter-systems provides the following relation between the system-output $\vartheta_i(r,p)$ and the source-function Φ_{Qi} as well as the boundary-function Φ_{Ri}, assuming boundary conditions of the Cauchy-type:

$$\vartheta_i(r,p) = \int_0^1 G_i(r,\eta,p)\Phi_{Qi}(\eta,p)d\eta + G_i(r,1,p)\Phi_{Ri}(p) \tag{2.14}$$

$G_i(r,\eta,p)$ is the Green's function of the system. It can be determined by comparison of equ. (2.13) and (2.14):

$$G_i(r,\eta,p) = \sum_{m=1}^{\infty} \frac{1}{k_{im}+p} \alpha_{im}\eta C_0(\gamma_{im}\eta) C_0(\gamma_{im}r)\varsigma_i c_i \tag{2.15}$$

We introduce the following radial coordinates:

η: radial coordinate of control-effectors
ς: radial coordinate of temperature-measurement
r: radial coordinate of controlled temperature

We assume that the essentially controlled temperature exists at one defined value of r, whereas temperature-measurement may be distributed throughout the whole system (Fig. 2) as well as the effector-actuation blood-flow (j = 1), metabolism (j = 2) and evaporation (j = 3). Environmental temperature shall be considered as the essential disturbance for this instance.

The Green's functions represent transfer-functions. $G_{ji}(r,\eta,p)$ indicates the influence of the component j of the source-function at the points η of the cylinder i on the controlled variable at the point r, and in the same manner $R_{ji}(r,\wp,p)$ the influence of the temperature deviation at the points $\wp$ on the effector-signal Y_j. Y_j actuates via local distribution-functions $c_{ji}(\eta)$ the source- and boundary-functions.

$$\Phi_{Qi}(\eta) = c_{1i}(\eta)\,Y_1 \qquad + c_{2i}(\eta)\,Y_2 \qquad + c_{4i}(\eta)\,Y_4$$

$$(2.16)$$

vasomotor-control	meta-bolism	blood-temperature

$$\Phi_{Ri} = c_{3i}\,Y_3 \qquad + \qquad c_{5i}\,Z_5 \qquad (2.17)$$

evaporation	air-temperature

The effector-signal is composed of three components, each of which is computed from the weighted and integrated temperature-deviations.

$$(2.18)$$

$$y_j = -\sum_{i=1}^{n}\int_{o}^{1} \Phi_i(\wp)R_{ji}(r,\wp)d\wp \qquad j = 1,2,3$$

In formal analogy to the Green's-functions for blood-flow, metabolism and evaporation R_{1i}, R_{2i}, R_{3i} one gets a fourth function R_{4i} for the central blood-temperature which is part of source-function. R_{4i} is composed of data of the passive system and of a term containing R_{1i}.

The solution of the whole system of equations should not be expanded in detail here. The result is obtained in form of a matrix-system. The elements of the coefficients consists of double integrals over η and $\wp$.

The transfer-behaviour of the controller is given by Green's functions $R_{ji}(r,\wp,p)$. As the thermoreceptors are proportionally and differentially reacting elements and additional neuronal lags have to

be considered, the dynamics of the controller should be described by
the following Green's-function:

(2.19)

$$R_J \nu(r, \wp, p) = V_J(r, \wp) g_J \nu(r, \wp) \; \frac{1 + \tau_1 \nu(r, \wp) p}{[1 + \tau_2 \nu(r, \wp) p][1 + \tau_3 \nu(r, \wp) p]}$$

V_J are the gains concerning the transformation of temperature
deviations into controller-actuations, $\tau_{1 \ldots 3} \nu$ are the time-constants
of the transferprocess and $g_J \nu$ weighting factors characterizing either the local distribution or the importance of receptors.

Using this model, for the first time radial dependencies of temperature within the body were computed (Werner 48, 53). As an example
Figs. 3 (A) and (B) show the dynamics of the radial profiles of the
trunk after a stepwise change of environmental temperature for a
naked man at rest (30 % relative humidity, 0.1 ms^{-1} air velocity).

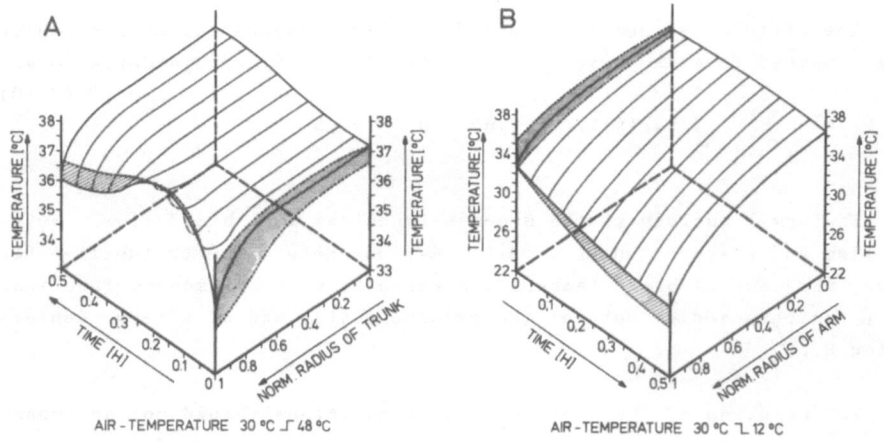

Fig. 3   Dynamics of radial temperature profiles in man obtained by
         mathematical simulation.
         Dashed lines: Experimental values. (A) Trunk, heat load.
         (B) Arm, cold load.

The step response to heat load (Fig. 3 A) is characterized by an overshoot in certain areas. In the direct centers (radial coordinate r = 0) we observe small overshoots only in the hands and the feet. The increase of temperature in the new steady-state is greater in the peripheral parts of the body, so that on the whole we obtain the effect of a local temperature equalization. In the head and the trunk (Fig. 3 A) we recognize a strong oscillation, as relatively low stationary skin temperatures are finally reached. In these parts there is a strong radial gradient of temperature only in the peripheral areas, while temperature at r = 0 increases gradually. A comparison between the final temperatures of the different parts of the body at the same radius yields only very small differences: at r = 0 about 0.5 °C and r = 1 about 2 °C. The maximum value at temperature difference in the final profiles is smaller than 2.5 °C.

The dotted and shaded areas show the range of experimental results as far as these can be obtained in man. The dotted areas reproduce radial profiles measured by Reader and Whyte (37). The results were taken from different parts of the trunk (Fig. 3 A) and arm (Fig. 3 B) and reveal that besides the radial coordinate, which here is the only one taken into account, particularly the axial coordinate and partly also the angular coordinate around the axes are of importance. The upper dashed boundary lines refer to the upper parts of trunk and arm, the lower lines to the more distal parts. Being aware that apart from this there are differences between the individuals and certain variations also in one single subject from experiment to experiment, the simulation results seem to fit quite well to reality. This holds also for the dynamics, where the dashed lines show our own experimental results (Werner, 50; Werner and Reents, 51) indicating that the final decrease of peripheral temperatures after heat load (Fig. 3 A) seems to be slower than extrapolated by the simulation, whereas after cold load it seems to be a bit faster (Fig. 3 B). However, repeated experiments show that the absolute temperature values may vary very much, whereas the relations with respect to the topography are rather constant. On the whole the experiments assure the correctness of the mathematical concept in principle and show agreement at least with the essential characteristics of the dynamics in experiment and simulation. After cold load the final steady-state is reached with partly enormous temperature decrease after a rather long time and without oscillations. Only the central areas of head, trunk, arm (Fig. 3 B) and leg reveal relatively small temperature decreases. In spite of the strong cold load we get rather a good constancy of the central parts of the body which of course involves temperature decrease of the peripheral areas, especially of hands and feet. In total, the process reveals temperature differences within the body of up to 15 °C.

3. Distributed Parameter Model: Three Dimensional Dependencies

In order to get the complete temperature and effector fields and to analyze the true distributed parameter control strategy it is necessary that the following requirements are fulfilled by the model:

a) All variables (e.g. temperature, heat-flow etc.) have to be regarded as functions of time and of three-dimensional local coordinates within the human body.
b) All parameters (e.g. density, conductivity index etc.) have to be considered as locally distributed parameters.
c) Geometry and anatomy of the body have to be represented adequately. This has been achieved by photogrametric analysis of anatomic models.
d) As heat transport mechanisms conduction, convection and radiation have to be taken into account separately.
e) As locally dependent effector mechanisms heat production by metabolism, vasomotor control and heat loss via sweat production and respiration have to be considered.
f) As disturbances to the control process environmental temperature, humidity, air velocity, eventually radiation and work load have to be incorporated.
g) The local definition of the really controlled variable (i.e. the temperature to be held as constant as possible) has to be adaptable to future results. Control of a single discrete temperature is as unprobable as control of complete temperature profiles.

The Ruhr-University has at its disposal a vector-computer CYBER 205, which is one of the fastest computers available. Therefore it was possible to develop a simulation program which is able to calculate on the basis of a three-dimensional digital representation of the human body temperature-profiles for various thermal loads. By giving up the (unrealistic) cylindrical symmetry, single organs and body areas may be identified. With the aid of this program it is furthermore possible to evaluate the effects of inhomogeneity and anatomy on a model of the body which corresponds closely to reality. On such a basis it should become possible to differentiate the sources of spatially distributed effects, which may be both the consequences of the properties of the passive system and of a distributed controller strategy. After the validation of this system various controller-models with lumped or distributed parameters could be implemented and evaluated. With these aims in mind, interaction with other systems is at present only taken into account with regard to heat loss via respiration and increase of muscle blood flow, cardiac output and tidal volume with increasing metabolic rate after cold or work load.

3.1 Data Base

A data base of the macroscopic anatomy of the human body has been developed by photogrammetric treatment of anatomical models (Somso Comp., AS 23, NS 9-15) and the further analysis of the thereby obtained sectional maps (Kelterbaum and co-workers, 29).
The gobal data are:

height	1.76 m
weight	67 kg
density	1.07 kg dm^{-3}
surface	1.82 m^2
volume	62.7 cm^3

For computer treatment the model was projected on a three-dimensional grid, for which in the head and the extremities 0.55 cm and in the trunk 1.1 cm were chosen. This makes possible a representation of vessels with a diameter ≥ 1.1 and ≥ 0.55 cm and of all essential organs of heat production. A higher resolution would have been desirable at many sites, e.g. in the hands and feet or in other parts of the body for a more precise modeling of the circulatory system. Although computer tomography would make possible such a data base, a simulation would go beyond the computing capacity of the available computers. With the aid of anatomical atlases (43, 28) and data on the distribution of tissue (45, 41, 14) and on the density of tissue (26) a satisfactory digital representation of the body has been obtained. Fig. 4 shows six examples of transverse trunk sections from the digital atlas.
Different symbols represent different tissues, which in the whole body add up to 54 different types, to which five parameters are

attributed: heat conductivity, specific heat capacity, density, basal
specific metabolic heat production, specific blood flow. For the skin
areas heat transfer coefficients α and basal and maximal sweat rate
SWR have been compiled.

As the actual version of the vector computer CYBER 205 at the Ruhr-
University did not allow processing as a whole, the data base was
divided into five segments which implies considerable additional data
transfer. The approximation of the body by cubes evokes a significantly increased body surface, which can principally not be eliminated by higher resolution, but is corrected by form factors.

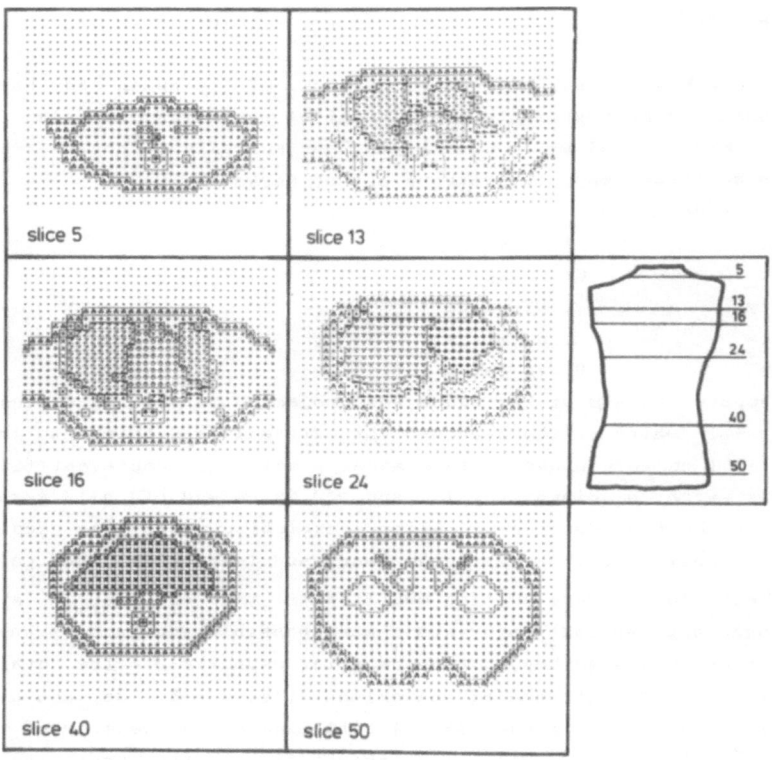

Fig. 4   Examples of transverse sections from the digital atlas for
          the trunk.

154

As true surface areas the values from DuBois und DuBois (16) have
been assumed. The handling of the data base is achieved by a separate
program which is independent of the special application. This makes
possible a quick adaptation of other atlases if even more efficient
computers were to become available.

3.2 Physics and mathematics of the passive system
3.2.2. Metabolic heat production

The dependence of biochemical reactions on temperature T is
described by the Q_{10}-effect: A doubling of reaction is achieved by a
temperature increase of 10 °C. Metabolic rate M in the passive system
is thereby a function of time t and the three-dimensional local
coordinate ξ (M_0 = metabolic rate at 37 °C)

$$M(\xi,t) = M_0(\xi,t)\, 2.0^{((T(\xi,t)-37.0)/10)} \tag{3.1}$$

Conductive heat transport
Assuming that conductive heat transport is isotropic, heat flow $\dot{q}_c$
is given by Fourier's law with heat conductivity λ

$$\dot{q}_c(\xi,t) = \frac{\partial}{\partial\xi}\left(-\lambda(\xi)\,\frac{\partial T(\xi,t)}{\partial\xi}\right) \tag{3.2}$$

At the tissue boundaries heat conductivity might change abruptly
which would evoke enormous additional problems for numerical solution. Therefore it was ascertained by use of a two-layer model
with stepwise λ-changes, that a linear approximation of the λ-step
brings about deviations in the temperature profiles which are smaller
than 1 %.

3.2.3 Convective heat transport

Convective heat transfer is determined by the geometry of the
vessels and the hydromechanical properties in the vessels. With the
aid of a simple model which assumes constant temperature and velocity
within a vessel's cross-section, one can estimate the amount of heat
transferred to the adjacent tissue, and which length of vessel is
needed for a complete balance between blood and tissue temperature.
Such a computation shows that in the big vessel there is hardly any
heat transfer with the tissue whereas in the capillaries there is
almost immediately a thermal balance with the environment. The thermoregulatory relevant heat transfer is obvious in the terminal arte-

ries, arterioles and venules. An exact simulation of vessels of this order is only possible for special cases and for models of single vessel networks (17). The vessels explicitly taken into account in this simulation exhibit only a small amount of heat transfer, however, the local temperatures can be considerably influenced by such vessels because they represent heat sources.

In accordance with the results reported in (10) we conclude that a description of convective heat transfer $\dot{q}_b$ for vessels of medium size is possible by the bio-heat approach which has also been used by Pennes (36), Wissler (59, 61), Stolwijk and Hardy (46) and Werner (50):

$$\dot{q}_b (\xi,t) = BF(\xi,t) \ \rho_b \ c_b \ (T_b - T (\xi,t)) \tag{3.3}$$

with blood flow BF, density ρ_b, specific heat of blood c_b and blood temperature T_b. The studies (28, 49) also confirm such considerations. In addition they showed that the counter-current principle shortens the efficient vessel distance for the balance of temperature with the adjacent tissue to about one third. As this distance for the vessels, taken into account in our simulation, is far longer than their length, it is justified to assume a spatially independent blood temperature within these vessels. It should be stated that such an approach is not able to describe anisotropy of convective heat transport (2, 65) nor to describe satisfactorily temperature fields in the vicinity of large vessels.

For the computation of the global time-dependent blood temperature T_b in the passive system an additional heat balance equation is necessary (α_b = heat transfer coefficient between vessel and tissue, m_b = mass of blood):

$$c_b \ m_b \ \frac{\partial T_b (\xi,t)}{\partial t} = \sum (BF(\xi,t) \ \rho_b \ c_b \ (T (\xi,t) - T_b)) \tag{3.4}$$
$$+ \sum \alpha_b (\xi) \ (T_b (t) - T(\xi,t))$$

Heat transfer at the skin

Heat transfer via radiation $\dot{Q}_r$ is according to the Stefan-Boltzmann-law:

$$\dot{Q}_r = \varepsilon \sigma \ A_r (T_{sk}^4 - T_r^4) \tag{3.5}$$

with Stefan-Boltzmann factor σ, emissivity index $\varepsilon \approx 0.95$, effective radiative surface A_r (for review see 24), skin temperature T_{sk} and radiative temperature T_r, both measured in K.

156

Conductive heat transfer $\dot{Q}_c$ is proportional to the difference of skin temperature T_{sk} and environmental temperature T_A:

$$\dot{Q}_c = \alpha_c \, A_c \, (T_{sk} - T_A) \tag{3.6}$$

with conductive heat transfer coefficient α_c. Convective heat transfer can be described by an equation analogous to Equ. (3.6) with a convective coefficient α_b which is dependent on air velocity v

$$\alpha_b = 2.7 + 7.4 \, v^{0.67} \text{ in } Wm^{-2} \, {}^\circ C^{-1} \tag{3.7}$$

which is based on an approach by Colin and Houdas (11) and Nishi (35). Heat transfer via evaporation $\dot{Q}_e$ is computed as the product of evaporative rate ER and evaporative enthalpie Δh_v of water vapor.

The maximal evaporation to the environment is proportional to the difference of water vapor pressure p for skin (sk) and air temperature (A). Saturated vapor pressure is non-linearly dependent on temperature. The relationship might be approximated by a fourth-order polynomial. Thus we get with skin surface A_e

$$\dot{Q}_e = \alpha_e \, A_e \, (p_{sk} - p_A) \tag{3.8}$$

The evaporative transfer coefficient may be computed by (for review see 12, 9, 24):

$$\alpha_e = 0.0448 + 0.123 \, v^{0.67} \text{ in } Wm^{-2} \, Pa^{-1} \tag{3.9}$$

In total we obtain a heat balance as boundary condition at the skin

$$- \lambda \, \frac{\partial T}{\partial n} = \frac{\dot{Q}_r}{A_r} + \frac{\dot{Q}_c}{A_c} + \frac{\dot{Q}_b}{A_b} + \frac{\dot{Q}_e}{A_e} \tag{3.10}$$

with the direction vector n, normal to the skin surface.

3.3 Controller Equations

According to Werner (55, 56) the steady-state of the thermoregulatory system is determined by the balance of the passive and active subsystems, i.e. of the controlled and controlling components. This balance is defined by the open-loop properties of the subsystems, as they are described for the passive system in Equ. (3.1-10). Such a concept implies minimal requirements for the biological realisation of the controlling system. In particular it dispenses with reference-signals or - temperatures. The mathematical description of the controlling system is intentionally very general, in order that various controlling concepts may be implemented and tested. Such a concept must fulfil the following requirements:

(1) each point of the body may be a receptor site
(2) processing in the contoller must make possible any combination of all receptor signals

157

(3) processing of receptor signals must take into account kind and site of effector mechanism

(4) time delays in the afferent, the processing and the efferent systems must be taken into account.

The dynamic properties of spatially distributed receptors are described by Equ. (3.11), where ζ denotes the site of receptor and τ_{r1}, τ_{r2} time constants. Via weighting functions f_{1i}, f_{2i} receptor signals y_r are processed to effector signals y_{ei}, η denoting the site of effector and i the effector system (i=1 metabolism, i=2 vasomotor action, i=3 sweat production):

$$y_r(\zeta,t) = \tau_{r1}(\zeta)T(\zeta,t) + \tau_{r2}(\zeta)\ \partial T(\zeta,t)/\partial t \qquad (3.11)$$

$$y_{ei}(\eta,t) = g_i(\eta,t)\quad [y_r(\zeta,t-t_0)\ f_{1i}(\zeta,\eta)$$

$$+ \int f_{2i}(\zeta,\vartheta,\eta)\ y_r(\zeta,t-t_0)\ y_r(\vartheta,t-t_0)\ d\vartheta]\ d\zeta \qquad (3.12)$$

$$+ y_{ei0}(\eta,t)$$

with gains g_i and basic effector signaly y_{ei0}.

A maximum simplification of these equations leads to controller equations with only two of three resultant afferent signals (e.g. core temperature, mean skin temperature, local skin temperature), which were proposed for metabolic heat production M by Stolwijk and Hardy (44, 47, 32), Bleichert and co-workers (6), for skin blood flow BF by Stolwijk and Hardy (47) and for sweating rate SWR by Wyndham and Atkins (66), Stolwijk and Hardy (44, 47) and Nadel (33, 34).

With regard to computer capacity the most complex practical realisation of the coupling matrix corresponding to the numerical representation of Equ. (3.12), may in the present simulation program contain 1024 elements for control of skin blood flow and sweating rate and 640 elements for control of metabolic heat production.

3.4 Numerical Solution

The passive system is essentially described by a three-dimensional diffusion equation with coefficients as functions of the local coordinates x, y, z. Such equations are usually solved by finite difference methods, and especially by a modification of the alternating direction implicit method, first reported by Douglas and Rachford (15).

For two reasons this method seemed to be advantageous:
1. It is an implicit method: the difference equation involved is
   stable for all quantities of Δx, Δy, Δz, Δt.
2. By partitioning one equation with three implicit directions into
   three equations with one implicit direction each, an algebraic
   reformulation results in three tridiagonal systems which can easily be solved by Gaussian algorithm.
The three tridiagonal systems are solved with the controller equations by turns.

3.5 Results of Simulation
3.5.1 Distribution of temperatures and profiles in neutral
environment

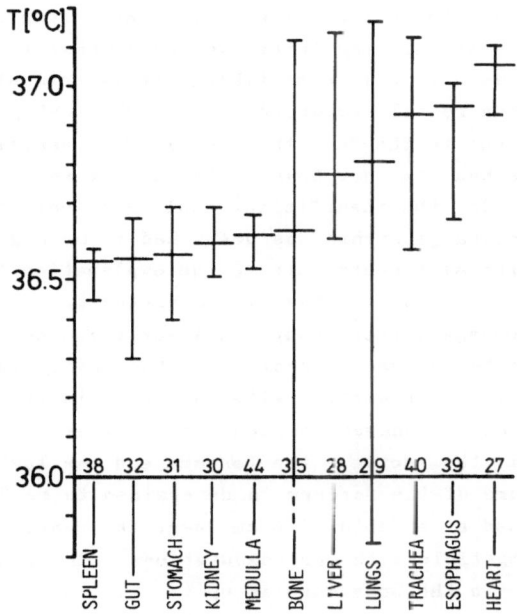

Fig. 5   Spatial span of various central temperatures of the trunk
          for neutral environment (mean values ± span).

In order to validate the model of the passive system, temperature profiles of the passive system in a neutral environment ($T_A=30°$) were computed. Metabolic and blood flow rates were chosen according to physiological knowledge. It is well known that temperatures in arms and calves are about 2 °C lower than in head, trunk and thighs, but the simulation reveals that even under neutral conditions these central temperatures vary by about 1 °C in the trunk (36.6 °C ±0.5 °C) and by 1.2 °C in the head (36.2 °C ±0.6 °C). Temperature of head muscles, presumably very near to tympanal temperature, is about 36.1 °C and thus is about 0.4 °C lower than intestinal temperature. Fig. 5 illustrates the spatial temperature span in various central tissues of the trunk under neutral conditions.

The maximal spatial temperature span is evident in the skeleton and the lung, but other organs, too, show considerable spatial temperature deviations even in indifferent conditions. Regarding the three experimentally used "core temperatures", namely tympanal, rectal and esophageal temperature, the latter is the highest in the simulation (37 °C). This is not fully compatible with experiments where it is either lower than rectal temperature or only slightly higher. This result may be due to the fact that cooling by ventilation of the neighboring trachea is not taken into account in the simulation. Another factor is the insufficient cooling effect of the heart, because heat exchange within the heart had to be neglected in the model. This might also contribute to the explanation for differences in experimental values of esophageal temperatures. With a different depth of the esophageal thermocouple a lower value could be evoked at a more distal site because of cooling by the respiratory tract and a higher value at a more central site in the vicinity of the heart. Aschoff and Wever (1) showed temperature profiles in three areas of the body, namely the rectum, the forearm and the lumbal region. The inner temperature of the forearm is determined to be 35.5 °C. This maximum is reached at a depth of 4 cm, near the center of the forearm and is fully compatible with our computations. The maximal temperature difference in the body core is up to +2.5 °C (at the sternum as compared to the rectum), other central temperatures measured at different sites differ by ±0.4 °C as compared to rectal temperature. Comparable results have been demonstrated by Eichna and co-workers (see 4) for temperatures in various vessels compared to the temperature of the Arteria femoralis. The computed temperature at sites which are also represented in the models of Stolwijk and Hardy (54, 47), Werner (51), Konz and co-workers (30), show good compatibility

with one another. Further experimentally determined temperatures were
presented by Pennes (36) for the arms, by Bazett and McGlone (5) for
subcutaneous regions, by Werner and Reents (54) and Werner and co-
workers (57) for the topography of skin temperatures. As compared to
the latter studies the present model slightly underestimates hand
temperatures, whereas other temperatures tend to be slightly over-
estimated. However, it has to be taken into account that such a
comparison has to be done by using computed values which are averaged
spatially for a body area, as the measuring sites in the literature
have not been provided with the exactness necessary for a point to
point comparison. Furthermore interindividual differences cannot be
overlooked as demonstrated by Wagner and Horvath (48): Group mean
values for different age and sex differ by ±0.2 °C for rectal and by
±0.5 °C for mean skin temperature. Also all reported experimental
results have hardly been obtained under totally comparable conditions. Air velocity, humidity and radiative temperature (!) vary
considerably, and there is no unequivocal consensus about neutral
environmental temperature. Many other factors like circadian rhythm,
psychic influences, adaptational status etc. contribute to many variations so that differences between experimental and computational
results may not only be attributed to deficencies of the model. In
conclusion we think that the present computation simulates a possible
mean test subject sufficiently well for us to dare to compute temperature profiles which cannot be obtained experimentally, this being
virtually one of the most worth-while features of simulation.

In the head temperature is extremely constant. Decrease to skin
temperature starts only in the skull. This is due to the high metabolic and blood flow rate of the brain. The profiles of the extremities
are, due to the lower rate more convex, but equally rather homogeneous and smooth. In the trunk especially the transverse profiles
depend very much on the longitudinal coordinate.
    Fig. 6 A-D elucidates the axial variation of transverse profiles
on account of geometry and inhomogeneity of the body.

The site of the sections can be read from the insets or from Fig. 3.
In section 5 the vicinity of the shoulder surface (a) has a strong
influence on the inner temperatures. This area is characterized by a
small volume to surface ratio implying lower local temperatures. In
the center (b) of this profile temperatures are considerably higher
($\approx$ 30 °C). The temperature profile shown in Fig. 6 B shows interesting local specialities, as e.g. the effect of high metabolic and

blood flow rates in the spinal cord. These are the reason for the moderate temperature decrease (c), as the adjacent tissue has a lower blood flow rate making possible a higher temperature. Above the heart (d) we recognize a moderate temperature increase. Heart temperature is about 37.4 °C and thereby 0.8 °C above rectal temperature. This is due to the high metabolic heat production of the heart. This effect will be retained both in a warm and cold environment. Section 24 (Fig. 6 C) shows a slightly lower temperature in the liver (e) compared to adjacent muscles. Temperature of the stomach (g) is even lower on account of the relatively low basal metabolic rate. The strong convexity of the temperature profile as an effect of lower skin temperature leads to a mean muscle temperature lower than stomach temperatures. Section 50 (Fig. 6 D) already has a bimodal temperature profile (i) indicating the neighbouring transition to the thighs. Effects of organs with high metabolic and blood flow rate are no longer visible.

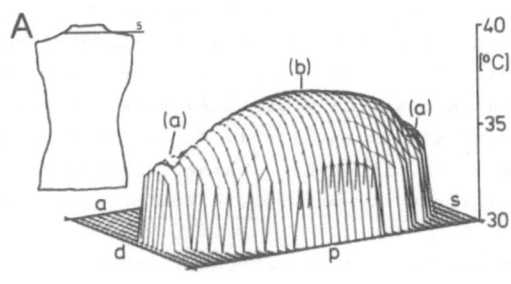

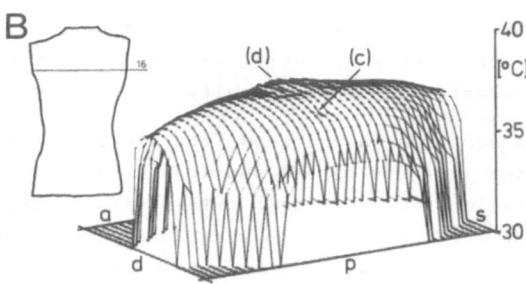

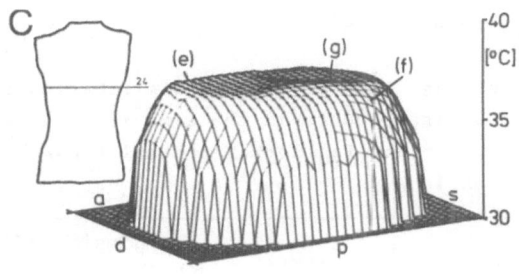

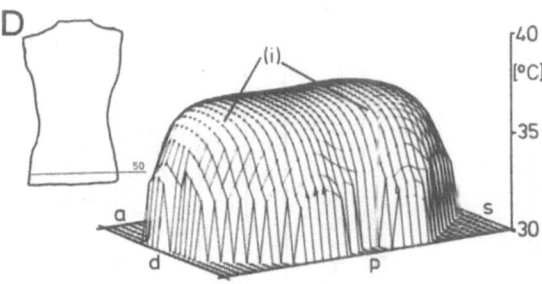

Fig.   6   Transverse temperature profiles in the trunk for neutral
A - D   environment. p = posterior a = anterior   d = dexter
              s = sinister. The sections are seen from caudal to cranial
              (site see inset). Indices (a) - (i) see text.

3.5.2 Control strategy

   An unsolved problem is the control strategy of the system, that is
the question whether the inhomogeneous pattern of effector distribution is maintained in the cold and warmth or whether there is
distributed active modification and control.
Therefore the lumped parameter approach delivering the best results
was first determined.
As controller input the following equation seems to be a reasonable
simplification

$$Y = 0.8\ T_{co} + 0.1\ T_{sk} + 0.1\ T_{mu} - 37 \qquad\qquad (3.13)$$

A correct global metabolic heat production M in Wm⁻³ may be computed
by $$M = -5000\,Y + 648. \qquad (3.14)$$

However, maintaining the pattern of distribution of heat production yields too high amounts in the arms and legs which do not
contribute much to the stabilization of central trunk and head temperatures (Fig. 7, example A).

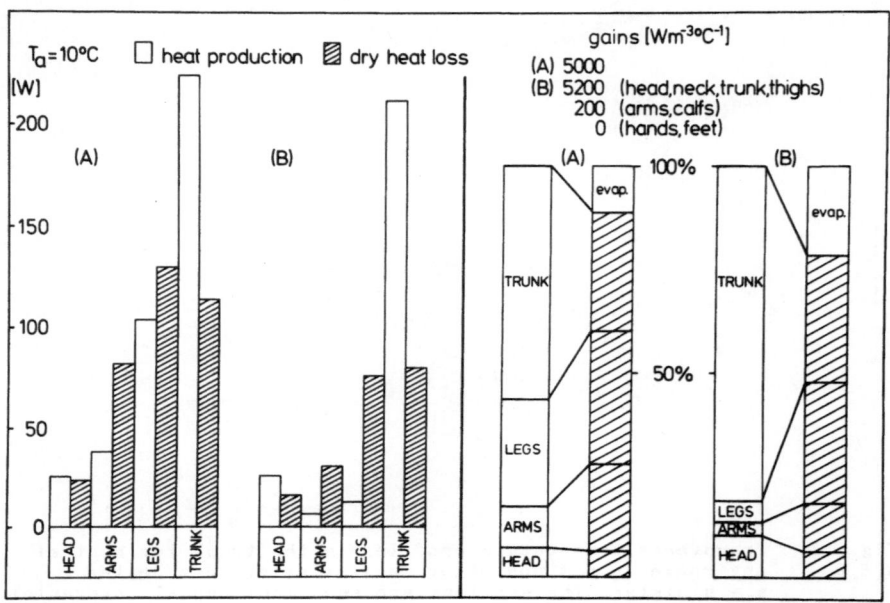

Fig. 7   Distribution of heat loss and heat production according to
          lumped parameter control (A) and distributed parameter (B)
          control of skeletal muscles. Air temperature 30 °C.

This may only be achieved by concentration of heat production in the
proximal areas and by reduction of dry heat loss (Fig. 7, example B).
Example B of Fig. 7 shows, however, by relative graphic presentation,
that evaporative heat loss via skin and respiration increases
slightly. Thus for cold defence a locally distributed control of heat
production in the skeletal muscles is required. The minimal requirements of a distributed control strategy are the following:

$$M = -5200\,Y + 648 \text{ for head and trunk}$$
$$M = -\ 200\,Y + 648 \text{ for arm and legs} \qquad (3.15)$$
$$M = \qquad\quad 648 \text{ for hands and feet.}$$

A similar but minor relevance of the distributed control concept has been confirmed with regard to blood flow rate. However, the influence of distributed minimal rates is more relevant than locally differing gains. As compared to the following quasi-optimal lumped control for blood flow BF in $m^3_{blood}\, m^{-3}_{tissue}\, s^{-1}$:

$$BF = 0.01\ Y + 0.0001, \qquad\qquad (3.16)$$

the following simplest distributed strategy delivers more realistic results:

$$BF = 0.01\quad Y + 0.0003 \text{ for forehead, nose, mouth}$$
$$BF = 0.01\quad Y + 0.0001 \text{ for remainder of head,}$$
$$\text{upper arm, thigh}$$
$$BF = 0.008\ Y + 0.0001 \text{ for trunk}$$
$$BF = 0.012\ Y + 0.0001 \text{ for forearm, calf} \qquad (3.17)$$
$$BF = 0.012\ Y + 0.0003 \text{ for hand, foot.}$$

On the other hand, the influence of distributed controller structures is not so obvious in warm defence with regard to evaporative heat loss, as there are very small temperature differences in the body after stronger warm load. The lumped approach for the sweating rate SWR in $gm^{-2}s^{-1}$

$$SWR = 0.1\ Y + Y_{min} \qquad\qquad (3.18)$$
$$(Y_{min} \text{ different for 16 areas)}$$

yields results which are very near to the following quasi-optimal distributed approach

$$SWR = 0.12\ Y + Y_{min} \text{ for forehead, nose, mouth}$$
$$SWR = 0.08\ Y + Y_{min} \text{ for remaining skin} \qquad (3.19)$$
$$Y_{min} \text{ different for 16 areas:}$$
$$177 \text{ (back of hand)} \ldots. 1766 \text{ (palm)}$$

Therefore for sweating rate, it cannot definitely be decided whether the distributed control is present or not.

3.5.3 Temperature profiles in the cold and in the warmth

Using the distributed approaches outlined above, longitudinal profiles for head, arm, leg and trunk were computed for various ambient temperatures as shown in Fig. 8.

Central profiles in arms and legs decrease enormously in the cold. At $T_A = 35\ ^oC$ and $40\ ^oC$, central temperatures of the extremities are nearly constant and very near to central temperatures of trunk and head. At $T_A = 10\ ^oC$ and $20\ ^oC$ the strong temperature decrease in the areas of the knee is due to the relatively high percentage of bone.

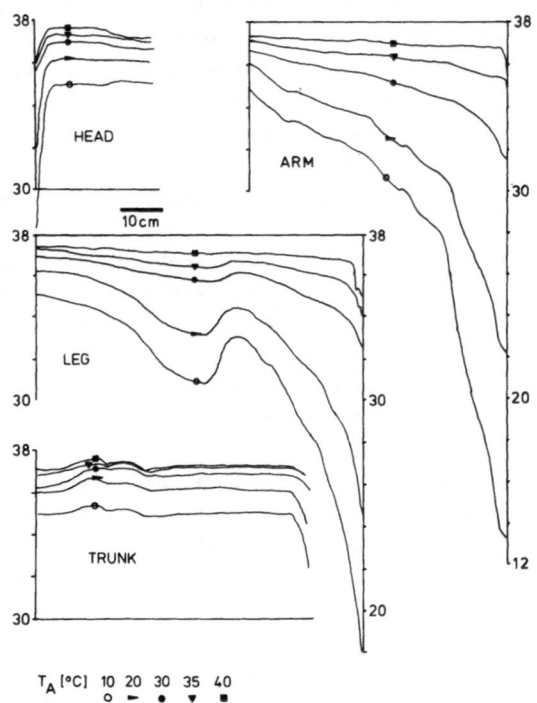

Fig. 8   Longitudinal profiles for various environmental temperatu-
          res TA.

Equally Fig. 9 demonstrates isotherms of the trunk in the level of
the collar bone.

A: 10°, B: 20°, C: 30°, D: 40° environmental temperature. A and B
demonstrate the reduction of the limited central warm areas,   whereas
at TA= 40 °C, temperatures lower than 36° are present only in the
areas of the shoulders.

The simulation of dynamic effects is possible but failed on
account of the small working storage (0.5 M words) of the CYBER 205
version at the Ruhr-University,   and inadequate changing frequency of
segments makes the computation inaccurate. Nevertheless first tests
show that the program is adequate for computation of dynamics as
well, and that the time courses will be computed correctly as soon as
the announced memory expansion becomes available.

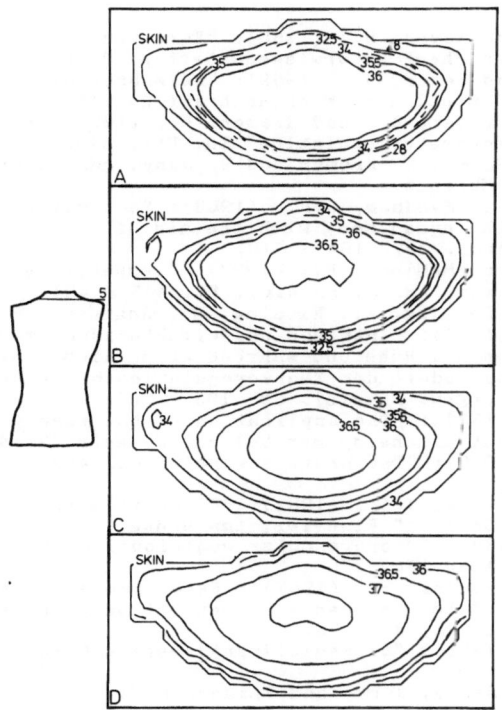

Fig. 9   Isotherms in the trunk at the level of the collar bone
          (see inset)
          for T_A = 10 °C (A), 20 °C (B), 30 °C (C) and 40 °C (D).

    With the availibility of more core capacity now, the future deve-
lopment must comprise the following features:

1. Local dependency of blood temperatures must be implemented.
2. Local coupling between sensor and effector sites has to be analyzed in more detail according to recent experiments (Heising
   and Werner (19, 20)).
3. In various areas of the body a higher resolution should be chosen.
4. A further coupling with other regulatory systems has to be taken
   into account.

References

1   Aschoff, J., and Wever, R., (1958): "Kern und Schale im Wärme-
    haushalt des Menschen" Naturwissenschaften 20, 477 - 487.
2   Aschoff, J., and Wever, R., (1959): "Wärmehaushalt mit Hilfe
    des Kreislaufs" Deut. med. Wochenschrift 84: 1509 - 1517.
3   Aschoff, J., Günther, B., and Kramer, K., (1971):"Energie-
    haushalt und Temperaturregulation" In: Physiologie des Menschen,
    vol. 2 (O.H. Gauer, K. Kramer and R. Jung, eds.), Urban &
    Schwarzenberg, München.
4   Atkins, A.R., and Wyndham, C.H., (1969): "A study of temperature
    regulation in the human body with the aid of an analogue compu-
    ter" Pflügers Arch. 307, 104 - 119.
5   Bazett, H.C., and McGlone, B., (1927): "Temperature gradients in
    the tissues in man" Am. J. Physiol. 82, 415 - 451.
6   Bleichert, A., Behling, K., Kitzing, M., Scarperi, M., and
    Scarperi, S., (1972): "Antriebe und effektorische Maßnahmen der
    Thermoregulation bei Ruhe und während körperlicher Arbeit.
    IV. Ein analoges Modell der Thermoregulation bei Ruhe und Ar-
    beit." Int. Z. angew. Physiol. 30, 193 - 206.
7   Burton, A.C., (1934): "The application of the theory of heat
    flow to the study of energy metabolism" J. Nutr. 7, 497 - 533.
8   Cabanac, M., (1975): "Temperature regulation" Ann. Rev. Phy-
    siol. 37, 415 - 439.
9   Cena, K., and Clark, J.A., (1981): "Bioengineering, Thermal
    Physiology and Comfort" Elsevier, Amsterdam.
10 Chen, M.M., (1980): "Microvascular contributions in tissue heat
    transfer" N.Y. Acad. Sc. 335.
11 Colin, J., and Houdas, Y., (1967): "Experimental determination
    of coefficient of heat exchange by convection of human body"
    J. Appl. Physiol. 22, 31 - 38.
12 Cooney, D.O., (1976): "Biomedical Engineering Principles"
    Dekker, New York.
13 Crosbie, R.J., Hardy, J.D., and Fessenden, E., (1963): "Elec-
    trical analog simulation of temperature regulation in man:
    Temperature, its Measurement and Control in Science and In-
    dustry, part 3 (D. Hardy, ed.), Reinhold, New York, 627 - 635.
14 Diem, K., and Lentner, C., (1968): "Documenta Geigy, Wissen-
    schaftliche Tabellen" Geigy, Basel.
15 Douglas, J. Jr., and Rachford, H.H. Jr., (1956): "On the nume-
    rical solution of heat conduction problems in two and three
    space variables" AMS 82, 421 - 439.
16 DuBois, D., and DuBois., E.F., (1915): "Clinical calorimetry 5:
    The measurement of surface area of man" Internal Med. 15, 868 -
    888.
17 Gross, J.F., and Popel, A., (eds.), (1981): "Mathematics of Mi-
    crocirculation Phenomena" Raven, New York.
18 Hardy, J.D., (1972): "Models of temperature regulation - a re-
    view" In: Essays on Temperature Regulation (J. Bligh and
    R. Moore, eds.), North-Holland, Amsterdam, 163 - 186.
19 Heising, M., and Werner, J., (1987): "Influences of overall
    thermal balance on local inputs for drive of evaporation in
    men" J. Appl. Physiol. 62, 926 - 931.
20 Heising, M., and Werner, J., (1987): "Control of sweating in
    man after work-induced thermal load and symmetrically applied
    cooling" Eur. J. Appl. Physiol. 56, 608 - 614.
21 Hellon, R.F., Townsend, Y., and Cranston, W.I., (1978):
    "A search for thermal receptors in central vasculature" In:
    New Trends in Thermal Physiology (Y. Houdas and J.D. Guieu),
    Masson, Paris, 101.
22 Hensel, H., (1952): "Physiologie der Thermoreception" Erg.
    Physiol. 47, 166 - 168.

23 Hensel, H., (1973): "Neuronal Processes in thermoregulation"
    Physiol. Rev. 53, 948 - 1017.
24 Houdas, Y., and Ring, E.F.J., (1982): "Human Body Temperature:
    its Measurement and Regulation" Plenum, New York.
25 Hwang, C.L., and Konz, S.A., (1977): "Engineering models of the
    human thermoregulatory system - a review" IEEE Trans. Biomed.
    Eng. 24, 309 - 325.
26 Jensen, R.K., (1978): "Estimation of the biochemical properties
    of three body types using a photogrammetric method" J. Biome-
    chanics 11, 349 - 358.
27 Jessen, C., Feistkorn G., Nagel A., (1983): "Temperature sensi-
    tivity of skeletal muscle in the conscious goat" J. Appl. Phy-
    siol. 54, 880 - 886.
28 Jiji, L.M., Weinbaum, S., and Lemons, D.E., (1984): "Theory and
    experiment for the effect of vascular microstructure on surface
    tissue heat transfer - part II: model formulation and solution"
    IEEE BME 106, 331 - 341.
29 Kelterbaum, J., Werner, J., and Schön, H., (1977): "Makroskopi-
    sche Topographie des menschlichen Körpers: Gewinnung der Rohda-
    ten und deren EDV-gerechte Aufarbeitung in einer Datenbank"
    EDV Med. Biol. 4, 123 - 128.
30 Konz, S., Hwang, C., Dhiman, B., Duncan, J., and Masud, A.,
    (1977): "An experimental validation of mathematical simulation
    of human thermoregulation" Comput. Biol. Med. 7, 71 - 82.
31 Kutznetz, L.H., (1979): "A two dimensional transient mathemati-
    cal model of human thermoregulation" Am. J. Physiol. 237, 266 -
    277.
32 Nadel, E.R., Horvath, S.M., and Dawson, A.T., (1970): "Sensiti-
    vity to central and peripheral thermal stimulation in man"
    J. Appl. Physiol. 29, 603 - 609.
33 Nadel, E.R., (1971): "Peripheral modifications to the central
    drive for sweating" J. Appl. Physiol. 31, 828 - 833.
34 Nadel, E.R., Bullard, R.W., and Stolwijk, J.A.J., (1971):
    "Importance of skin temperature in the regulation of sweating"
    J. Appl. Physiol. 31, 80 - 87.
35 Nishi, Y., (1981): "Measurement of thermal balance" In: Bio-
    engineering, Thermal Physiology and Comfort (K. Cena und
    J.A. Clark, eds.), Elsevier, Amsterdam, 29 - 40.
36 Pennes, H.H., (1948): "Analysis of tissue and arterial blood
    temperature in the resting human forearm" J. Appl. Physiol. 1,
    93 - 122.
37 Pernkopf, E., (1973): "Topographische Anatomie des Menschen"
    Urban & Schwarzenberg, München.
38 Rawson R.O., and Quick, K.P., (1970): "Evidence of deepbody
    thermoreceptor response to intra-abdominal heating of the ewe"
    J. Appl. Physiol. 28, 813 - 820.
39 Reader, S.R., and Whyte, H.M., (1951): "Tissue temperature gra-
    dients" J. Appl. Physiol. 4, 396 - 402.
40 Riedel, W., Siaplauras, G., and Simon, E., (1973): "Intra-abdo-
    minal thermosensitivity in the rabbit as compared with spinal
    thermosensitivity" Pflügers Arch. 340, 59 - 70.

41 Schmidt, R.F., and Thews, G., (1985): "Physiologie des Menschen"
    Springer, Berlin.
42 Simon, E., (1974): "Temperature regulation: The spinal cord as a
    site of extrahypothalamic thermoregulatory functions" Rev. Phy-
    siol. Biochem. Pharmac. 71, 1 - 76.
43 Sobotta, J., and Becher, H., (1972): "Atlas der Anatomie des
    Menschen" Urban & Schwarzenberg, München.
44 Stolwijk, J.A.J., and Hardy, J.D., (1966): "Temperature regula-
    tion in man - a theoretical study" Pflügers Arch. 291, 129 -
    162.

45 Stolwijk, J.A.J., and Hardy, J.D., (1966): "Partitional calori-
    metric studies of responses of man to thermal transients"
    J. Appl. Physiol. 21, 967 - 977.
46 Stolwijk, J.A.J., (1970): "Mathematical model of thermoregula-
    tion" In: Physiological and behavioral temperature regulation
    (J.D. Hardy, A.P. Gagge and J.A.J. Stolwijk, eds.), Thomas,
    Springfield, 703 - 721.
47 Stolwijk, J.A.J., and Hardy, J.D., (1977): "Control of body
    temperature" In: Handbook of Physiology 9, Am. Phys. Society,
    45 - 68.
48 Wagner, J.A., and Horvath, S.M., (1985): "Cardiovascular reac-
    tions to cold exposures differ with age and gender" J. Appl.
    Physiol. 58, 187 - 192.
49 Weinbaum, S., Jiji, L.M., and Lemons, D.E., (1984): "Theory and
    experiment for the effect of vascular microstructure on surface
    tissue heat transfer - part I: Anatomical foundation and model
    conceptualization" IEEE BME 106, 321 - 330.
50 Werner, J., (1974): "Zur Dynamik der menschlichen Temperaturre-
    gulation" Habilitationsschrift, Ruhr-Universität Bochum.
51 Werner, J., (1975): "Zur Temperaturregelung des menschlichen
    Körpers. Ein mathematisches Modell mit verteilten Parametern und
    ortsabhängigen Variablen" Biol. Cybern. 17, 53 - 63.
52 Werner, J., (1981): "Control aspects of human temperature regu-
    lation" Automatica 17, 351 - 362.
53 Werner, J., (1977): "Mathematical treatment of structure and
    function of the human thermoregulatory system" Biol. Cybern.
    25, 93 - 101.
54 Werner, J., and Reents, T., (1980): "A contribution to the topo-
    graphy of temperature regulation in man" Eur. J. Appl. Physiol.
    45, 87 - 94.
55 Werner, J., (1980): "The concept of regulation for human body
    temperature" J. Therm. Biol. 5, 77 - 82.
56 Werner, J., (1984): "Regelung der menschlichen Körpertempera-
    tur" de Gruyter, Berlin - New York, 288 pp.
57 Werner, J., Heising, M., Rautenberg, W., and Leimann, K.,
    (1985): "Dynamics and topography of human temperature regulation
    in response to thermal and work load" Eur. J. Appl. Physiol. 53,
    353 - 358.
58 Werner, J., (1986): "Do black-box models of thermoregulation
    still have any research value? Contribution of system-theoreti-
    cal models to the analysis of thermoregulation" Yale J. Biol.
    Med. 59, 335 - 348.
59 Wissler, E.H., (1961): "Steady state temperature distribution in
    man" J. Appl. Physiol. 16, 734 - 740.
60 Wissler, E.H., (1963): "An analysis of factors affecting tempe-
    rature levels in the nude human" In: Temperature - its Measure-
    ment and Control in Science and Industry (J.D. Hardy, ed.),
    Reinhold, New York, 603 - 612.
61 Wissler, E.H., (1964): "A mathematical model of human thermal
    system" Bull. Math. Biophys. 26, 147 - 166.
62 Wissler, E.H., (1970): "The use of finite difference techniques
    in simulating the human thermal system" In: Physiological and
    Behavioral Temperature Regulation (J.D. Hardy, A.P. Gagge and
    J.A.J. Stolwijk, eds.), Thomas, Springfield, 367 - 388.
63 Wissler, E.H., (1985): "Mathematical simulation of human thermal
    behavior using whole body models" In: Heat Transfer in Medicine
    and Biology (A. Shitzer and R.C. Eberhart, eds.), Plenum,
    New York, 325 - 374.
64 Wissler, E.H., (1988): "Modeling human exposure to thermal
    stress" In: Environmental Ergonomics (I. Mekjavic, ed., in
    press).

65 Wulff, W., (1980): Discussion paper "Alternatives to bioheat transfer equation" Ann. N.Y. Acad. Sc. 335, 151 – 154.
66 Wyndham, C.H., and Atkins, A.R., (1968): "A physiological scheme and mathematical model of temperature regulation in man" Pflügers Arch. 303, 14 – 30.

Supported by Deutsche Forschungsgemeinschaft, SFB 114 + We 919/2-1.

Large-scale Multiple Model for the Simulation of Anesthesia

R.Q.Y. Tham, F.J. Sasse and V.C. Rideout

Department of Electrical and Computer Engineering
and Department of Anesthesiology
University of Wisconsin-Madison, WI 53706

INTRODUCTION

During anesthesia, many interacting factors (both intrinsic and extrinsic to the physiological system) affect hemodynamic responses and equilibria. Cardiovascular parameters, which are intrinsic factors of the physiological system regulating the hemodynamic responses, constantly adjust to maintain metabolic demand. External factors such as drug actions exert their influence directly on the cardiovascular parameters, and indirectly through the baroreceptor, chemoreceptor, and hormonal regulators to alter the cardiovascular parameter values and hemodynamic responses. Interactions among the drug actions and the cardiovascular system are complex and varied, producing different responses between and within species. One example of such variations arises with the multiple sites of actions of halothane as identified by Seagard, *et al.* [Seag85], accounting for some inconsistent observations in heart rate and systemic peripheral resistances responses that occur during halothane anesthesia.

To analytically describe physiological behaviors during anesthesia, mathematical models of various complexity, which interrelate drug concentrations and arterial pressures, have been devised. On the very simplest scale, linear models with lead-lag and dead-time transfer functions have been used to describe the relationships between anesthetic dosages as input and the mean arterial pressure as output. These models have been used in control schemes in servo-anesthesia control systems, for example [Schi87]. Often, the time constants and delays are assumed not to change with the cardiovascular dynamics and levels of anesthesia. However, these simple models simulate the complex physiological behaviors under anesthesia inadequately.

To add realism and physiological basis to the anesthesia model, complex large-scale multiple models detailing the cardiovascular dynamics and the transport of anesthetic agent to the individual tissue beds are required. Such large-scale anesthesia models with well-defined design goals and sufficient structural details integrate the available experimental information to explicitly describe the interactions of anesthetic agents with the multiple sites of physiological actions [Fuku81a,81b]. With more realistic representation of the controlled physiological system than simpler models, large-scale models pose better challenges in the simulation testing of control schemes under design, and give insights into the physiological factors that influence the major time constants in the simpler lead-lag anesthesia models.

A physiologically-based multiple anesthesia model (a mass transport driven by a baroreceptor controlled non-pulsatile pressure-flow model) describing the interactions of cardiovascular circulation and baroreceptor control with halothane is introduced as an example of a large-scale anesthesia model. The convenience of individually setting the potency and dynamics of drug effects at various sites of actions permits sensitivity analyses of the drug action at various sites, and comparisons of local effects to the total effect of the drug on the cardiovascular responses. Digital computer simulation of the mathematical model reproduced some typical and atypical hemodynamic responses that are observed during halothane anesthesia. Explanations for some of these atypical responses were revealed by examining the effects of basal physiological states (prior to changes in anesthetic levels), pharmacokinetics, potency and dynamics of halothane at various sites of actions. This physiologically-based model can also serve as a basis for the development of computer-aided programs to teach dose behaviors of anesthetic agents.

CONCEPT OF THE LARGE-SCALE MULTIPLE MODEL

Mapleson [Mapl73] was one of the early pioneers who modelled halothane anesthesia with transport in various tissue beds. In his linear human model, halothane was distributed to various tissue beds according their partition coefficients and perfusion. The rate of gas distribution was quantized into multiples of unit dead-time delays simulating transit times of blood circulation in the major blood vessels and tissue beds. Blood circulation was unaffected by halothane concentration. The canine version of this model was vadidated by experimental data [Allo76]. Ashman, *et al.* [Ashm70] developed the first closed-circulation halothane uptake and distribution transport model driven by blood circulation that was linearly dependent upon the arterial and venous halothane concentration. This continuous model consisted of two compartments in the respiratory uptake submodel and six compartments in the cardiovascular transport submodel (representing the viscera and endocrine, brain, muscle/skin, fat, connective tissues, and lung). An analog computer simulated this human model. Many papers on physiologically based modelling in pharmacokinetics by Dedrick, Bischoff and others began to appear at about the same time [Dedr68].

Realistically, halothane exerts different dose-dependent effects on several cardiovascular parameters (such as heart rate and contractility, vascular resistances in various tissue beds, large venous compliances and unstressed blood volumes). Changes in the cardiovascular parameter values belonging to the tissue beds depend upon the rates of halothane distribution and, possibly, on its actions. To simulate these changes, a more complex physiologically-based multiple model as introduced by Beneken and Rideout is needed [Bene68, Ride75]. The original approach uses pulsatile blood flows in a pressure-flow (P-F) model to drive linear mass transport models. An early use of multiple model to simulate halothane anesthesia was developed by Zwart, et al. [Zwar72]. In addition to having more compartments in the distribution submodel than Ashman's anesthesia model, tissue blood flows and vascular resistances, simulated by the P-F model, were individually and linearly changed as they were affected by the concentrations of halothane in the corresponding tissue beds.

The most comprehensive application of multiple modelling in halothane anesthesia was developed by Fukui, et al. [Fuku81a, Fuku81b], based in part on his earlier model-based studies [Fuku73]. Their multiple model consisted of two compartments in the respiratory model and eight tissue compartments in the cardiovascular circulation. They extended the original concept of multiple modelling beyond the P-F and transport model combination to incorporate reflex control, direct halothane influences, and a parallel carbon dioxide transport submodel, all of which interact with each other. The pulsing ventricles, the baroreceptor and chemoreceptor parametric feedback loops, and the dependence of cardiovascular parameters on halothane distribution contributed to a non-linear, time-varying multiple model. Again, this human model was implemented in an analog computer.

MODEL DESIGN

COMPOSITION OF THE ANESTHESIA MODEL

Figure 1 shows the decomposition of the large-scale anesthesia model used in this study. It consist of a respiratory and a cardiovascular submodel. The respiratory model is completely represented by the respiratory multiple model which describes the halothane transport between tidal ventilation and blood. This respiratory multiple model includes a simple pressure-flow and a mass transport model.

The cardiovascular submodel, which is the principal model of interest, is functionally

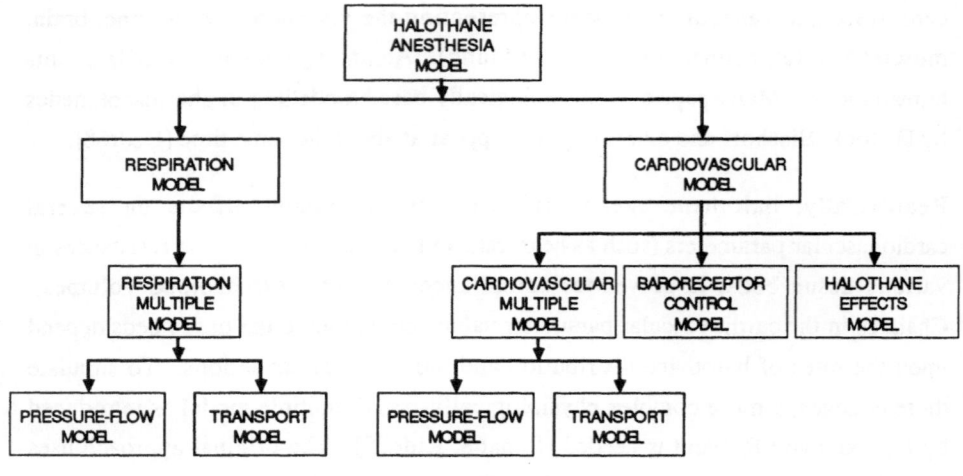

Figure 1. Decomposition of the Halothane Anesthesia Model.
Structural Relationships are not Shown.

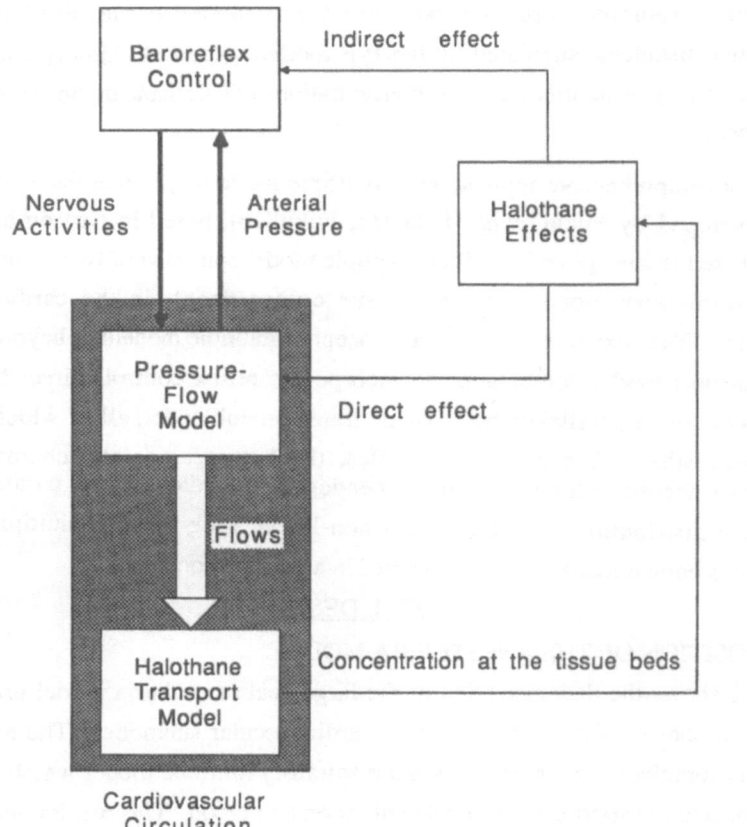

Figure 2. Structural Overview of the Cardiovascular Submodel

decomposed into the hemodynamic, baroreflex, pharmacokinetic, and pharmacodynamic submodels. Figure 2 shows the interactions of the components within the cardiovascular submodel. Several interacting loops operate simultaneously. The blood circulation affects the rates of halothane distribution to various tissue beds, which in turn alter the rate of change of cardiovascular parameter values that govern blood circulation. This interaction is further complicated by the inclusion of the baroreflex control which regulates blood pressure and is also affected by halothane.

The core of the cardiovascular submodel is comprised of the P-F and the mass transport submodels which together form the cardiovascular multiple submodel. Topologically, the cardiovascular multiple model (see fig. 3) mimics a closed circulation with blood flowing from the right heart into the pulmonary system, left heart, and returning through the systemic circulation. Elements of the cardiovascular multiple model include the left and right heart, arteries, veins, and tissue beds consisting of pulmonary, coronary, brain white and gray matter, muscle/skin, well and poorly perfused organs, and adipose tissue. These tissue beds are typically the ones chosen in large-scale anesthesia models, for example [Mapl73, Bedu79, Fuku81a].

HEMODYNAMIC SUBMODEL

The hemodynamic submodel is modelled by segments of blood vessels whose bloodflow is pumped by the left and right heart. Although pulsatility in the blood flow has been used in the multiple model studies, it can be omitted in models that have rates of anesthetic distribution and reactions that are much slower than the frequency of pulsatility. While this simplification may seem a regression from realism in the modelling of halothane anesthesia, the goals of this study and the advantages in computer runtimes of using a non-pulsatile blood flow model justify its use here. Furthermore, it is shown [Tham88] that the simulated mean blood pressures in the pulsatile and non-pulsatile anesthesia model respond almost identically to halothane anesthesia. By avoiding pulsatile pressure-flow waveforms, high frequency components in blood flows are eliminated. The model computes the slower and steady state responses according to the method described in [Ride83] and [Ride85].

In the hemodynamic submodel, segments of blood vessels are represented by vascular resistance and compliance. Their relationships can be written in equations related to electrical analogs, as shown in fig. 4, using lumped compliance, C_n, and resistance, R_n. Equations describing the n-th segment of the blood vessel which relate blood flows, pressures and volumes are as follows:

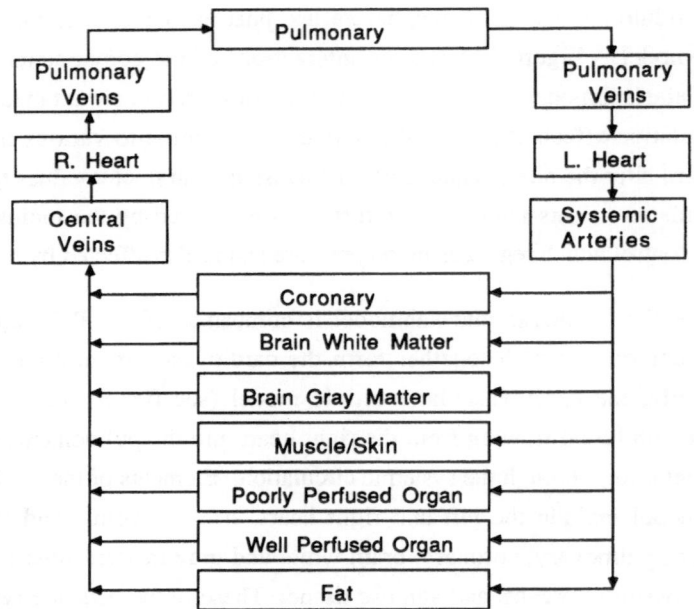

Figure 3.   Structure of the Cardiovascular
multiple model.

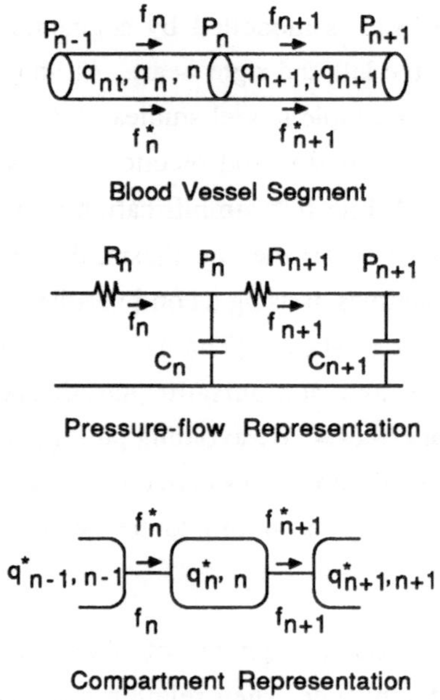

$$P_{n-1} \quad \xrightarrow{f_n} \quad P_n \quad \xrightarrow{f_{n+1}} \quad P_{n+1}$$

Blood Vessel Segment

Pressure-flow Representation

Compartment Representation

Figure 4.   Building Blocks of the Cardiovascular
Submodel

$$F_{n-1} - F_n \quad = dC_nP_n\,/\,dt \qquad\qquad\qquad\qquad \text{...1-a}$$
$$P_n - P_{n+1} \quad = R_{n+1}\,F_{n+1} \qquad\qquad\qquad\qquad \text{...1-b}$$
$$Q_{nt} \qquad\quad = P_n\,C_n + Q_{nu} \qquad\qquad\qquad\qquad \text{...1-c}$$

where P's, F's, and Q's are average blood pressures, flows and volumes, respectively, of the elements subscripted. Subscripts u and t denote the unstressed and total blood volumes of the subscripted element. Segments of blood vessels are linked together to form the vascular system and replicate the spatial distribution of the circulatory system. Vascular compliances of blood vessels in the systemic circulation are lumped together as systemic arterial or venous compliance. Applying equations 1-a to 1-c to the pressure-flow model, as shown in fig. 5, we obtain the following equations:

$$F_S - F_R \quad = dC_RP_R\,/\,dt$$
$$F_R - F_P \quad = dC_PP_P\,/\,dt$$
$$F_P - F_L \quad = dC_LP_L\,/\,dt$$
$$F_L - F_S \quad\; = dC_SP_S\,/\,dt$$
$$P_P - P_L \quad = R_PF_P \qquad\qquad\qquad\qquad\qquad\quad \text{...2}$$
$$P_S - P_R \quad = R_{CO}F_{CO} + R_{BG}F_{BG} + R_{BW}F_{BW}$$
$$\qquad\qquad\qquad + R_{WP}F_{WP} + R_{PP}F_{PP} + R_{MS}F_{MS} + R_{FA}F_{FA}$$
$$Q_R \qquad\;\; = Q_{RU} + C_RP_R$$
$$Q_P \qquad\;\; = Q_{PU} + C_PP_P$$
$$Q_L \qquad\;\; = Q_{LU} + C_LP_L$$
$$Q_S \qquad\;\; = Q_{SU} + C_SP_S \qquad\qquad\qquad\qquad\; \text{...2-a}$$
$$Q_{TOTAL} \; = Q_R + Q_P + Q_L + Q_S \qquad\qquad\qquad \text{...2-b}$$

It must be noted that, equation 2-b is an excess state equation. The total blood volume, Q_{total}, is not conserved if disturbances occur in the unstressed blood volume, vascular compliances, or pressures. To remove this dependency, one of the stressed volume equations, such as equation 2-a, must be omitted, and equation 2-b must be rewritten as

$$Q_S = Q_{TOTAL} - (Q_R + Q_P + Q_L) \qquad\qquad\qquad \text{...2-c.}$$

In the non-pulsatile P-F model proposed in [Ride83], equations describing cardiac outflow of the left and right ventricles are determined by the filling and pumping action of the heart. The single cycle of the ventricular pressure-volume state trajectory which traces a counter-clockwise locus is shown in fig. 6. Use of such state-space representation is based in part on the work of Sagawa and his colleagues

179

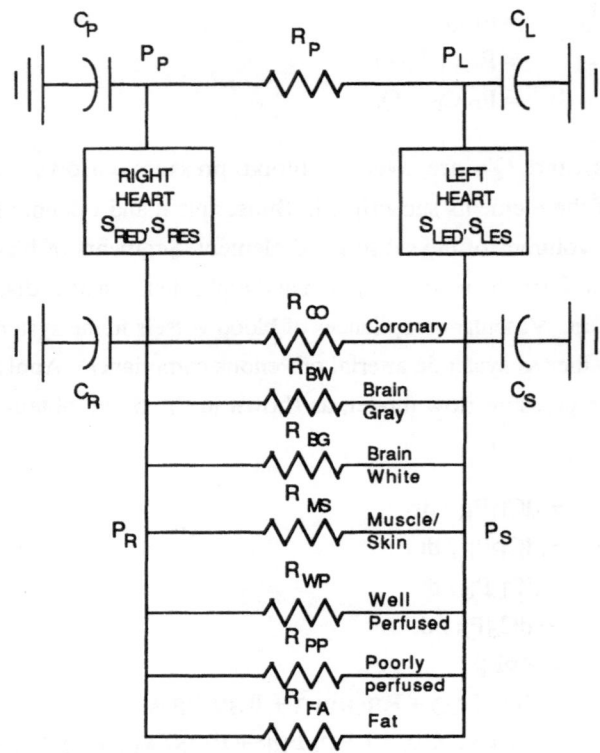

Figure 5.   Structure of the Non-pulsatile P-F Model.

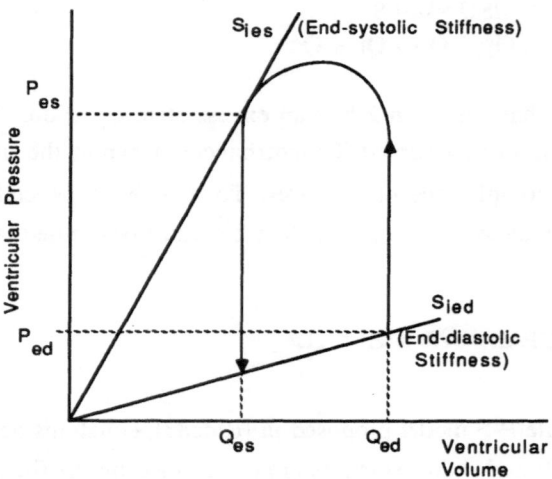

Figure 6.   Instantaneous Pressure-Volume Relationship

[Suna82]. The end-diastolic and end-systolic volumes are shown as the maximum and minimum ventricular blood volumes. Sagawa and colleagues showed that these extreme volumes can be determined by the end-diastolic (subscripted as ed) and end-systolic (subscripted as es) pressure-volume stiffnesses and pressures. Furthermore, for a given contractile state, the end-systolic pressure (and consequently, the stroke volume) depends only on the end-diastolic volume and not the time course of the state trajectory.

By definition, stroke volume is the difference between end-diastole and end-systole blood volumes (q_{ed} and q_{es} in fig. 6, respectively). Multiplying stroke volume by

$$
\begin{aligned}
\text{Cardiac output} \quad &= \text{HR} * \text{Stroke volume} \\
&= \text{HR} * (q_{ed} - q_{es}) \\
&= \text{HR} * (P_{ed} * S_{ed} - P_{es} * S_{es}) \qquad \ldots 3
\end{aligned}
$$

where P and S represent pressures and ventricular stiffness. This ignores possible non-linearities in the stiffness relationships, which may be attributed to (a) the tension property of cardiac muscle, (b) inadequate filling which tends to decrease effective diastolic stiffness, and (c) the geometry of the heart. In practice, the end-diastolic and end-systolic pressures are approximately equal to, or can be converted to the mean venous filling and mean afterload pressures by empirical conversion factors, k_{ed} and k_{es} [Ride85]. After post-division and pre-multiplication of the end-diastolic and end-systolic pressures in eq. 3 with the corresponding conversion factors and then regrouping and defining S_{pre} and S_{aft} as the preload and afterload pressure-volume stiffness, we can approximate the cardiac output in terms of the mean arterial and venous pressures as follows:

$$
\begin{aligned}
\text{Cardiac output} \quad &= \text{HR} * ((k_{ed} * P_{ed}/k_{ed}) * S_{ed} - k_{es} * P_{es}/k_{es}) * S_{es})) \\
&= \text{HR} * (P_{venous} * S_{pre} - P_{arterial} * S_{aft}) \qquad \ldots 4.
\end{aligned}
$$

In terms of the variables used in this model, cardiac output of the left and right heart (F_L and F_R, respectively) are written as

$$
\begin{aligned}
F_L \quad &= \text{HR} * (P_L * S_{Led} - P_S * S_{LES}) \\
F_R \quad &= \text{HR} * (P_R * S_{Red} - P_P * S_{RES}) \qquad \ldots 5
\end{aligned}
$$

At equilibrium, F_L and F_R are equal in magnitude. However, during halothane induction and other disturbances these blood flows may exhibit transient differences.

COMPARTMENTAL TRANSPORT MODEL
The development of cardiovascular transport models is well documented [Midd72].

Transport models are built using a finite number of non-distributive lumped compartments which may correspond to vessels or tissue beds which carry the substance under study. This assumption reduces the equations that represent each compartment from spatial-temporal partial differential equations to simple time domain ordinary differential equations. Some spatial distribution of the anesthetic transport model is preserved by the cascading of several mixing chambers. As the number of chambers used is finite the model response is band-limited as in the P-F model.

The basic transport equation of a compartment is derived from the law of conservation of mass, which gives a rate of change in concentration of the solute equal to the net integral of inflow and removal of the solute normalized by the effective volume of the compartment. The effective volume of the tissue compartment referred to blood is the product of the tissue/blood partition coefficient and the tissue volume. For this study, the concentration of halothane is measured by partial pressures. Differential equations describing the dilution of halothane are:

$$d(\gamma_n)/dt = (\Sigma \gamma_{in} F_{in} - \Sigma \gamma_n F_{out})/V_b \qquad \text{for blood vessels} \quad ...6\text{-}a$$
$$d(\gamma_n)/dt = (\Sigma \gamma_{in} F_{in} - \Sigma \gamma_n F_{out}))/(\lambda_n V_{nt}) \qquad \text{for tissue beds} \quad ...6\text{-}b$$

where γ_n and γ_{in} are halothane concentrations (referred to blood) of the nth compartment and mean blood flow (F_{in}) entering the same compartment; F_{out} is the mean blood flow leaving the compartment; V_b and V_{nt} are the blood and tissue volumes; and λ_n is the tissue/blood partition coefficients. If concentration are to be referred to air, the blood/gas partition coefficient, λ_b, must also be used.

The structure of the transport model is topologically similar to the cardiovascular model, as shown in fig.3. The corresponding basic transport equations are applied to each of the segments of blood vessels or tissue beds to obtain the equations describing the distribution of halothane concentration in the cardiovascular system. Blood flows entering and leaving each compartment are computed from the P-F model.

BARORECEPTOR CONTROL SUBMODEL

To achieve the goal of examining the indirect effects of halothane on the baroreflex, detailed non-linear representation of the baroreflex is needed. Detailed representations of baroreflex were attempted by Dick [Dick68], Katona [Kato80], and Martin [Levy79; Mart70]. The baroreflex model developed by Martin is extensively used in this anesthesia model. In this submodel, shown in fig. 7, the baroreflex is functionally divided into the afferent, central, and efferent submodels. The baroreflex control uses systemic arterial pressure as input to the reflex loop. Transduction from

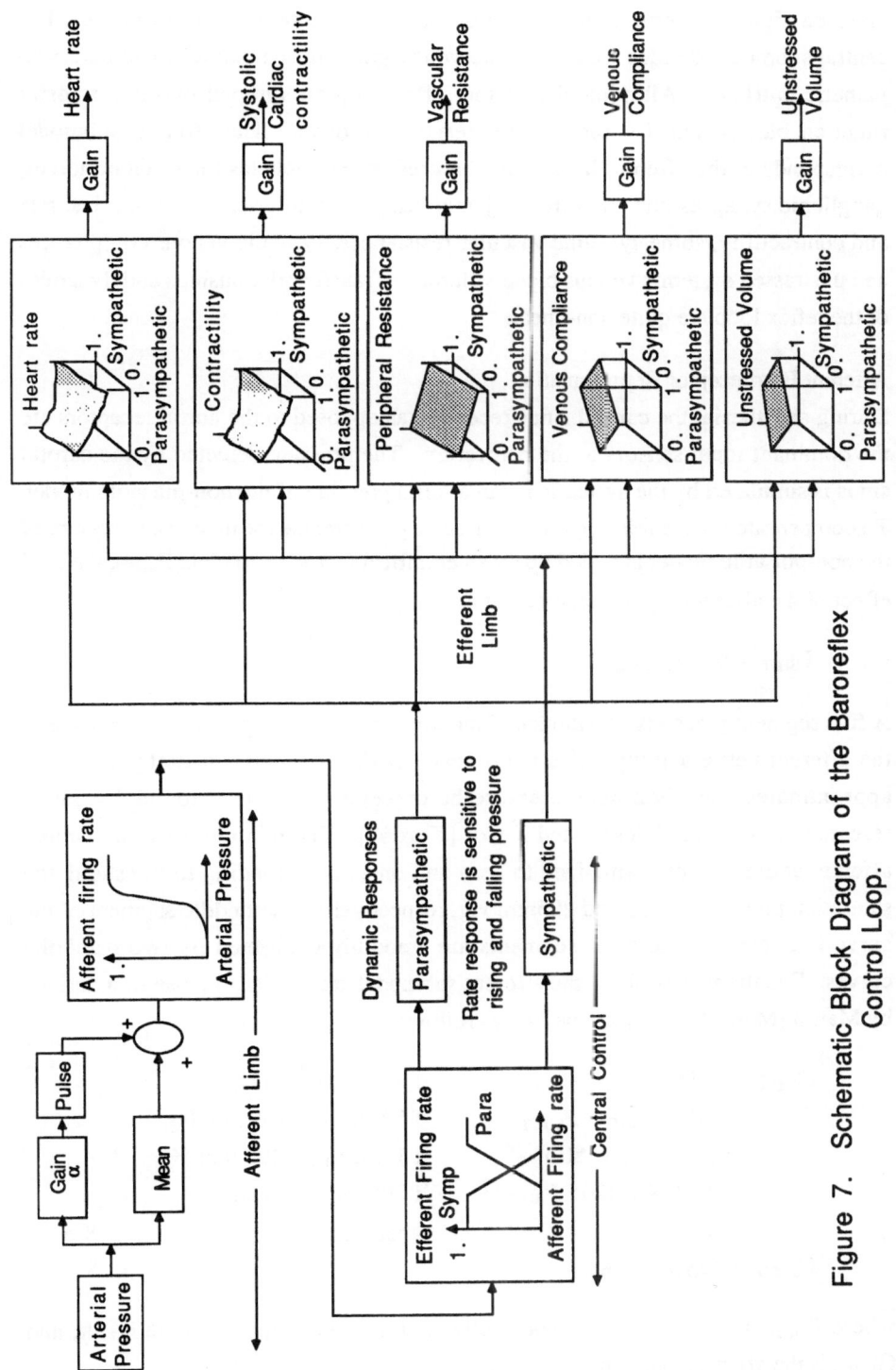

Figure 7. Schematic Block Diagram of the Baroreflex Control Loop.

183

pressure signal to nerve impulses takes place in the afferent control block. The central submodel divides the afferent nervous signal into sympathetic and parasympathetic outflows. All dynamics of the reflex loop are lumped into two transfer function blocks, one for each of the nervous outflows. The efferent submodel corresponds to the efferent limb of the baroreflex and includes the efferent nerves, ganglionic synapses and end-effector junctions. The baroreflex affects cardiac rate and contractility, some systemic vascular resistances, systemic venous compliances and unstressed systemic venous blood volumes. Transfer relationships and dynamics of the reflex loop are quite non-linear.

Afferent Baroreceptor Representation

During anesthesia, the carotid sinus receptors as opposed to the aortic receptors are the dominant input sensors to the baroreflex. The pressure detected by the carotid sinus is simulated by the systemic mean arterial pressure of the non-pulsatile model. To compensate for the missing effects of pulse pressure, the mean arterial pressure of the non-pulsatile model is scaled up by an empirical factor, α, to yield P_{sum}, the total effect of the afferent input pressure, that is

$$P_{sum} = \alpha * P_{arterial} \qquad \ldots 7.$$

A five segment piecewise-continuous function converts the input arterial pressure to the afferent nerve activity. The non-linear transfer function, illustrated in fig. 7, approximates the sigmoidal pressure-baroreceptor activities found by many researchers, such as Spickler and Kezdi [Kezd67]. The minimum and maximum afferent activities are normalized to zero and one, corresponding to threshold and saturation pressures of 50 and 205mm Hg, respectively. The middle segment of the curve is linear. These three segments are smoothly connected by two parabolic curves. Equations describing the afferent submodel are similar to those first derived by Martin [Mart70]. They are written as follows

$$
\begin{aligned}
F_{bar1} &= 0.0 & &\text{if } P_{sum} < 50 \text{ mm Hg} \\
&= 0.01 * P_{sum} * P_{sum} & &\text{if } 50 < P_{sum} < 105 \text{ mm Hg} \\
&= .01 * (P_{sum} - 80.) & &\text{if } 105 < P_{sum} < 205 \text{ mm Hg} \\
&= 1. - .0001 * (205 - P_{sum})^2 & &\text{if } 205 \text{ mm Hg} < P_{sum} \\
&= 1. & &\text{otherwise} & \ldots 8\text{-a} \\
F_{baro} &= G_{baro} * F_{bar1} & & & \ldots 8\text{-b}
\end{aligned}
$$

where F_{baro} is the afferent nervous activity; F_{bar1} is an intermediate result; and G_{baro} is the afferent gain which is sensitized by halothane [Seag83].

Central Baroreceptor Submodel

The central control submodel separates the afferent nervous activities into the sympathetic and parasympathetic nervous outflows. This models the brain segment of the baroreflex loop. Unfortunately, due to lack of quantitative data, the higher brain center modulation of the baroreflex cannot be modelled in much detail. For convenience, the time course of the entire baroreflex loop is lumped here. Although it is physiologically incorrect to ascribe the time course of the entire baroreflex to the brain segment (which is known to have a very small time delay and a negligible time constant [Mart70]) this formulation is structurally equivalent to inserting lead and lag networks in the afferent submodel and, repeatedly, in every efferent nerve activity.

Two three-segment piecewise continuous transfer functions in the central control translate the afferent signal to efferent outflows. At low afferent activities, sympathetic activity is saturated while parasympathetic activities are cut off. The converse occurs at high afferent activities. Straight lines describe the middle segment of the transfer functions. The equations below generate the sympathetic and parasympathetic activities, respectively.

$$
\begin{array}{lll}
F_{symp} & = 0. & \text{if } F_{baro} > 1. \\
& = F_{symp,th} - G_{symp}*F_{baro} & \text{if } 0. \leq F_{baro} \leq 1. \\
& = 1. & \text{if } F_{baro} < 0. \qquad \ldots 9\text{-}a
\end{array}
$$

$$
\begin{array}{lll}
F_{para} & = 0. & \text{if } F_{baro} < 0. \\
& = G_{para}*(F_{baro} - F_{para,Th}) & \text{if } 0. \leq F_{baro} \leq 1. \\
& = 1. & \text{if } F_{baro} > 1. \qquad \ldots 9\text{-}b.
\end{array}
$$

Each of the efferent branches has a second-order overdamped dynamic transfer characteristic, given (in Laplace transform) by

$$F_{out}(s)/F_{in}(s) = G_i / [(1+st_1)*(1+st_2)] \qquad \ldots 10$$

where $F_{out}(s)$ and $F_{in}(s)$ are the Laplace transform of the input and output nervous activities of the brain segment; s is the Laplace operator; G_i is the gain sensitivity of the efferent output that is dependent on halothane; and t_1, t_2 are time constants. Time constants are sensitive to the direction of pressure changes, but are unchanged by halothane anesthesia. Such dependencies are documented in [Kezd67] and [Mart70].

Efferent Baroreceptor Submodel

The parasympathetic and sympathetic activities simultaneously control the cardiac rate and contractility. Their relationships are described by ellipsoidal functions [Mart70]. Equations describing the control of heart rate are as follows:

$$F_{hsymp} = 2. * F_{symp} * Gpara_{hal}$$

$$F_{hpara} = 4. * F_{para} * Gsymp_{hal}$$

$$DHR = GHR * (19.64F_{hsymp} - 17.95F_{hpara} - 1.225F_{hsymp}*F_{hsymp} + 1.357F_{hpara}*F_{hpara} - 1.523F_{hsymp}*F_{hpara})$$

$$HR = HR_{intrinsic} * GHR_{hal} * (1+DHR) \qquad ...11$$

where F_{symp} and F_{para} are the efferent sympathetic and parasympathetic outputs of the central baroreflex submodel; $Gpara_{hal}$ and $Gsymp_{hal}$ are the gain sensitivities corresponding to ganglionic blockade of halothane; GHR is the barocontrol gain sensitivity of the heart rate; HR is the heart rate; DHR is the baroreflex modulation of the intrinsic heart rate; F_{hsymp} and F_{hpara} are the respective sympathetic and parasympathetic inputs to the the end-effector organ; GHR_{hal} is the direct effect gain sensitivity of halothane on the heart rate. Equations describing the control of cardiac contractility are as follows:

$$FS_{symp} = F_{symp} * Gpara_{hal}$$

$$FS_{para} = F_{para} * Gsymp_{hal}$$

$$DS_i = GS_i*(1.4 +17.6FS_{symp} -9.3FS_{para} -13.7FS_{symp}*FS_{symp} +.5FS_{para}*FS_{para} -3.35FS_{symp}*FS_{para})$$

$$S_i = S_{i,intrinsic}*GS_{hal} * (1.+DS_i) \qquad ...12$$

where S_i is the left or right end-systolic stiffness; DS_i is the baroreflex modulation of intrinsic end-systolic stiffness; $S_{i,intrinsic}$ is the denervated contractile state of the heart; FS_{symp} and FS_{para} are the respective sympathetic and parasympathetic inputs to the end-effector organ; GS_i is the barocontrol gain sensitivity of the end-systolic stiffness; GS_{hal} is the direct effect gain sensitivity of halothane on cardiac contractility.

In the heart rate equations (eqn. 11), the coefficients of the corresponding sympathetic and parasympathetic efferent terms are comparable, and thus may appear that the efferent activities are equally potent. However, the inputs to the ellipsoidal function are scaled such that parasympathetic activities are twice as large as the sympathetic activities [Mart70]. In eqn. 12, when the coefficients of the sympathetic and parasympathetic terms are compared, it is evident that the sympathetic nervous tone dominates the cardiac contractility, although contractility does become more sensitive to the parasympathetic activities when the sympathetic activities increase. The effect of the sympathetic activity on contractility is less dependent on the parasympathetic tone. These equations are in agreement with established experimental observations. In Martin's study, left ventricular systolic pressure determines the state of cardiac contractility. Here end-systolic pressure-volume ratio measures the state of contractility. Either measure will indicate the relative change in cardiac contractile strength. End-diastolic compliance is unaffected by the baroreflex.

Currently there is an absence of experimental data that simultaneously relates the efferent outflows to the resistances, compliances, and unstressed blood volumes of the peripheral vessels. Thus decoupled linear functions were assumed to describe the sympathetic and parasympathetic control of peripheral vessels. Here a weighted sum of the efferent outflow, F_{peri}, acts as the nervous input to the peripheral vessels. Parasympathetic activity, which is known to be less potent in the peripheral control of blood vessels [Guyt81], is weighted less than the sympathetic activities. The intrinsic properties of the blood vessels are then modified by this combined efferent nervous signal, F_{peri}. Equations describing each transfer block are

$$F_{peri} = F_{symp} - 0.2F_{para}$$
$$P_j = P_{j,intrinsic} * (1 + G_j * F_{peri}) \qquad ...13$$

where P_j is the parameter of the peripheral vessel affected by the baroreflex; G_j is the barocontrol gain sensitivity which can be negative [Tham88].

PHARMACODYNAMIC SUBMODEL

Although the halothane effects on the cardiovascular system have been closely examined, most studies tended to estimate the transfer gains between the anesthetic as input and the hemodynamic responses as output. Data describing the dynamic and gain characteristics of halothane at local sites of actions are rarely reported. The time course of the anesthetic effects are assumed to be contributed solely by the transport of halothane from the site of uptake to the sites of action. The lack of appropriate experimental data is a major obstacle to the development of large-scale models.

Direct Pharmacological Effects

The heart and systemic arteries are known be affected by halothane directly. These are modelled by dose-dependent functions that alter the intrinsic values of cardiac rate [Seag82] and contractility (as determined by measurements of guinea pig cardiac tissue tension, by [Lync81]. The intrinsic heart rate is assumed to vary linearly with halothane concentration. A quadratic relation is used to fit, by regression, the experimental data that described the effects of halothane on cardiac contractility. The following equations describe the effects

$$GHR_d = (1. - \gamma_{co} * KHR) \qquad ...14$$
$$GS_d = .96 - K_{co1}*\gamma_{co} + K_{co2}*\gamma_{co}*\gamma_{co}$$

where GHR_d, and GS_d are the direct gains modifying the intrinsic cardiac properties;

γ_{CO} represents the concentration of halothane in the myocardial circulation; K_{HR}, K_{CO1}, and K_{CO2} are parameters governing the effect of halothane. Halothane does not directly affect the peripheral vessels in the model.

Indirect Pharmacological Effects

In a very comprehensive review on the effects of halothane, Seagard [Seag82] reported that halothane sensitizes the baroreceptors, but blunts the central and efferent limbs of the baroreflex loop. This is modelled by modifying the transfer gains at the appropriate tissue sites. The transfer gains vary linearly with local drug concentration. The following equations describe their effects

$$
\begin{aligned}
G_{BARAR} &= 1. + K_{BARAR} * \gamma_{AR} \\
G_{BAROG} &= 1. - K_{BAROG} * \gamma_{BG} \\
G_{SYMEG} &= 1. - K_{SYMBG} * \gamma_{BG} \\
G_{PAREG} &= 1. - K_{PARBG} * \gamma_{BG} \\
G_{SYMCO} &= 0.986 - K_{SYMCO} * \gamma_{CO} \qquad \qquad \dots 15 \\
G_{PARCO} &= 0.969 - K_{PARCO} * \gamma_{CO} \\
G_{MS} &= 1. - K_{RHAL} * \gamma_{MS} \\
G_{WP} &= 1. - K_{RHAL} * \gamma_{WP}
\end{aligned}
$$

where G's are the static gains of the baroreceptors (subscripted as BARAR), brain (BAROG), sympathetic and parasympathetic ganglionic blockades (SYMEG, PAREG), end-effector synapses at the heart (SYMCO, PARCO), and muscle/skin (MS) and well-perfused organ tissues (WP); K's are empirical dose dependent gains;

SIMULATION RESULTS

The halothane anesthesia model was simulated using canine data on digital computers using a high level language, ACSL [Mitc86], which translates into FORTRAN for compilation. On the AT class of personal computer this program executed at two-thirds real time. Simulated responses were validated qualitatively and quantitatively, [Tham88]. Trends of the cardiovascular parameter changes and the blood pressure, flow and volume responses were facially examined for consistency by an anesthesiologist, and against accepted norms of physiological behaviors of the canine cardiovascular system under halothane anesthesia as observed in animal experiments [Tham88], and other similar large-scale models [Fuku81a]. A typical response of the model to a one percent step increase in halothane concentration is shown in fig. 8.

Other qualitative validation include sensitivity analyses of the mean arterial pressure

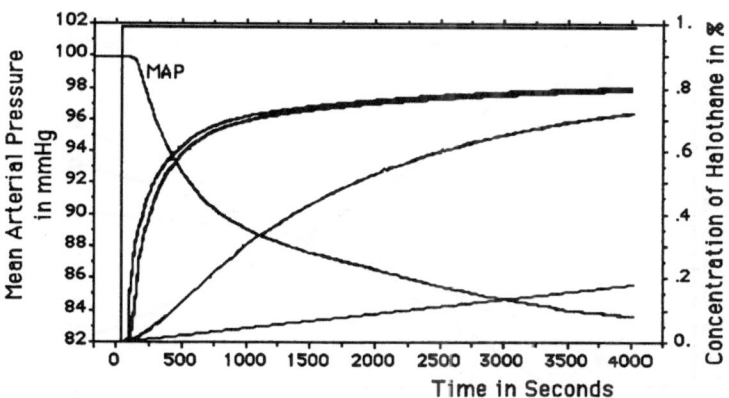

Figure 8. Transient Mean Arterial Pressure, Uptake and Distribution of 0.0 to 1.0% step increase in halothane. Halothane concentration curves from left to right are inspired, arterial, myocardial, poorly perfused organ and fat.

(with and without the baroreflex control) to variations of cardiovascular parameter values, see fig. 9. The sensitivity analyses were conducted at three simulated levels of halothane. Pressure response was most sensitive to to ventricular contractility at all levels of halothane, supporting the established explanation that halothane induced hypotension is of cardiac in origin. At equilibria mean arterial pressure responses (100, 95, 70 mm Hg) was non-linearly related to drug dosages (0.0, 0.5, 1.0% halothane, respectively). This behavior arises from the non-linear open-loop gains of the baroreflex, specifically the parasympathetic and quadratic heart rate gains. Higher halothane dosage increasingly blunts the baroreflex arc, reduces loop gains and causes parameter sensitvities to increase.

To show the ability of the large-scale model to explain physiological behaviors, the anomalous blood pressure response, observed by Schils [Schi83], was examined using the halothane model. In the anomalous blood pressure response, instead of a monotonic decrease in blood pressure following an increase in halothane anesthetic level (which depresses cardiac contractility and dilates blood vessels), blood pressure briefly rises before falling to its expected steady state. The converse response was also observed when halothane anesthetic levels were decreased. It was hypothesized, within the framework of the model, that the reversal in blood pressure can only be caused by the blockade of the parasympathetic activities by halothane. This was examined by simulating a step increase in halothane starting with a high basal parasympathetic activity in the canine model and a weak depressive halothane action on cardiac contractility. Contrary to established physiological observations, blood

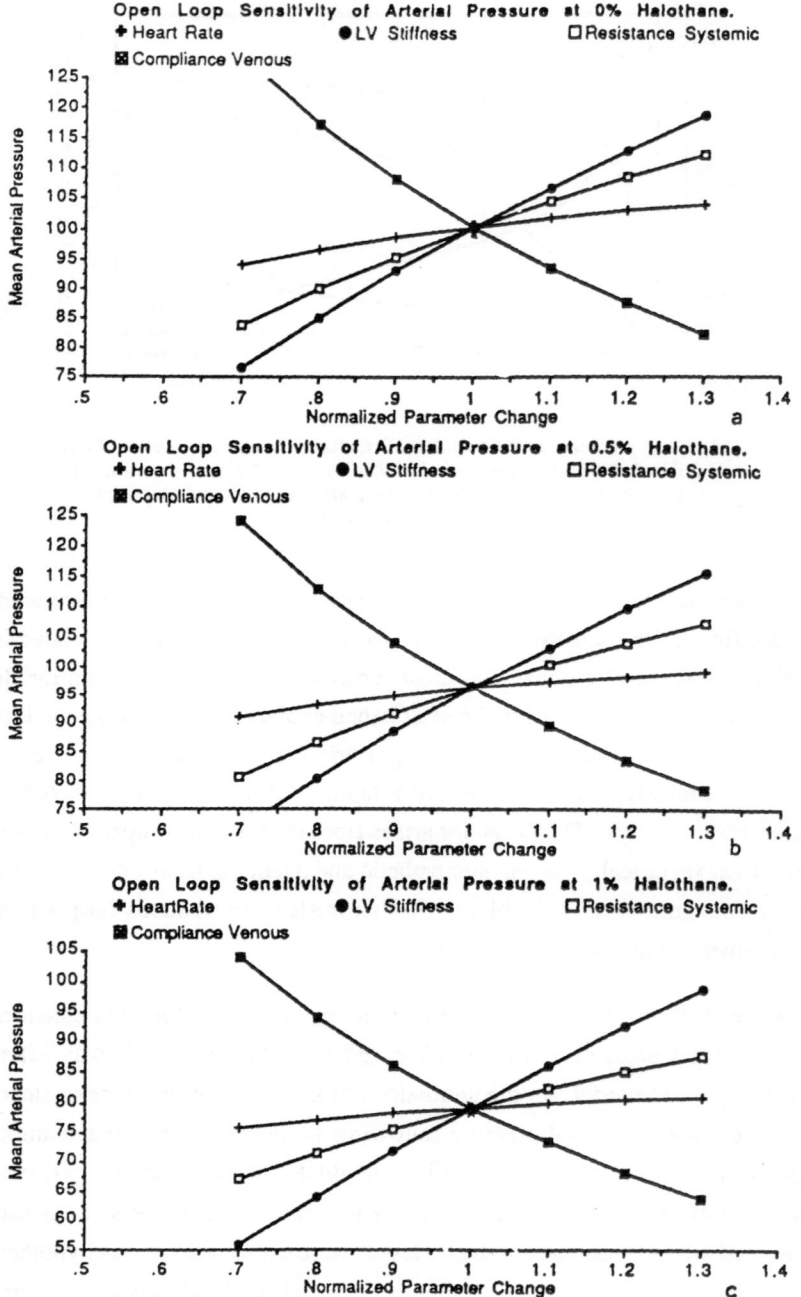

Figure 9a. Arterial Pressure Sensitivity to the cardiovascular parameters variations at different levels of halothane. The baroreflex loop was opened.

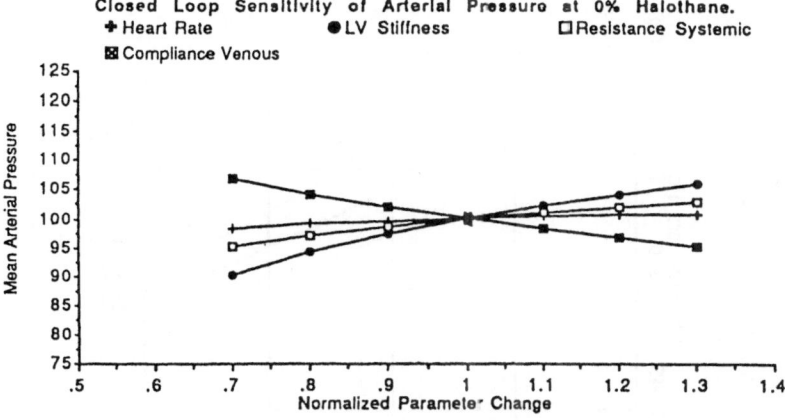

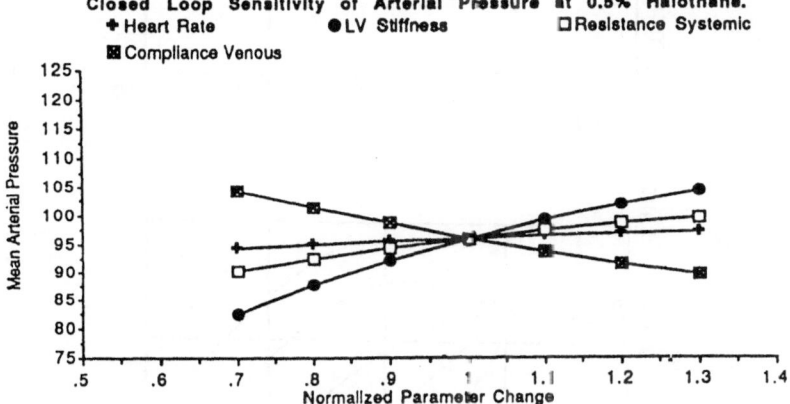

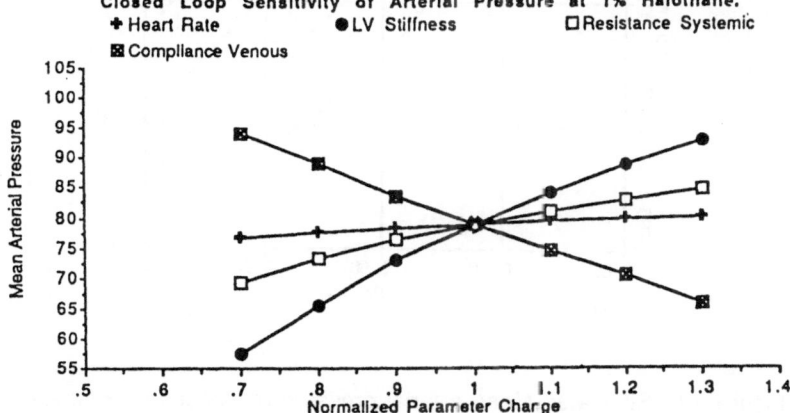

Figure 9b. Arterial Pressure Sensitivity to the cardiovascular parameters variations at different levels of halothane. The baroreflex loop was intact.

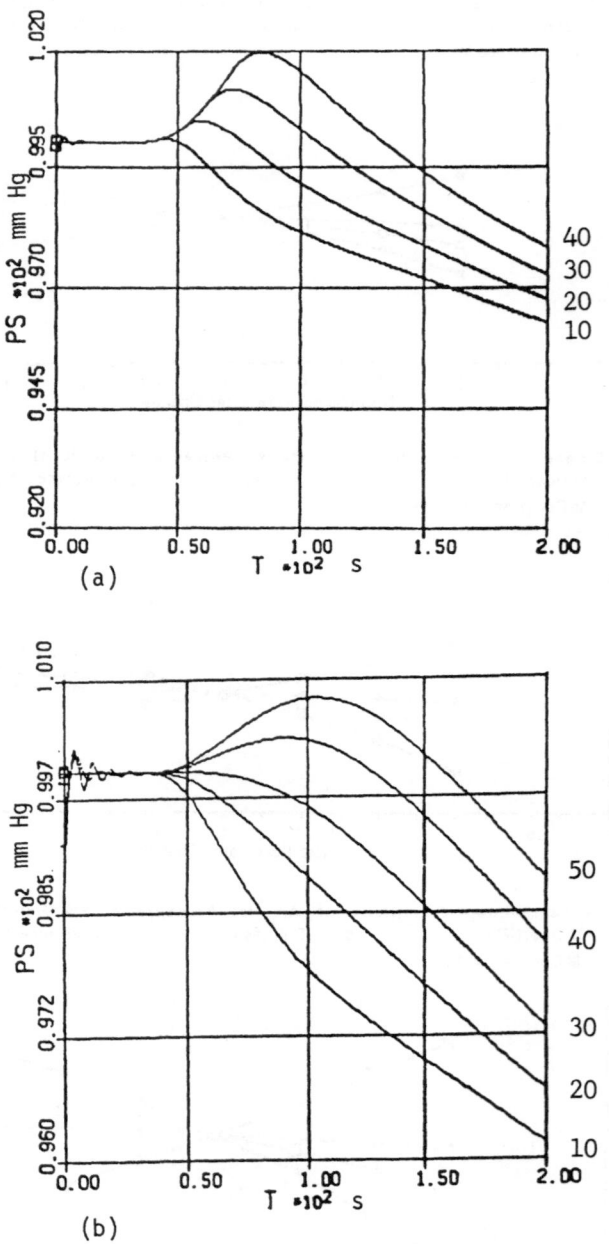

Figure 10. Simulated Mean arterial pressure responses to a canine anesthesia model with (a) 10 to 40 s of pure time delays and (b) 10 to 50 s of time lag in the onset of the direct effect of halothane to depress the cardiovascular parameters with respect to the parasympathetic blockade during a 0.0 to 1.0 % step increase in halothane.

pressure rose monotonically to a steady state - the anomalous transient response was not observed. However, the anomalous blood pressure response was observed if the depressive drug action was allowed to lag the parasympathetic blocking effect of halothane, (see fig. 10). This was simulated by adding time lags or pure time delays in the path giving the depressive effects of halothane. While this simulation excercise may reveal a deficiency in the initial halothane model in explaining the anomalous response, the results suggest that the transport of halothane by itself does not account for all the time response of blood pressure due to halothane. Reaction delays at local sites of action may contribute to the dynamics of blood pressure responses, and are often ignored by pharmacist and modellers. The presence of the dynamic action of halothane at local sites of action cannot be discovered by the use of simple lead-lag models.

CONCLUSIONS

This study examined the use of large-scale multiple model to simulate physiological responses during anesthesia. Submodels describing the hemodynamics, the barore-flex, the pharmacokinetics, and the pharmacodynamic behaviors and interactions were represented. The model was validated by experimental observations and model-to-model comparisons. Further, the study presented an explanation for an anomalous blood pressure response observed in [Schi83], demonstrating the usefulness of the model as an analytical and predictive tool. In the model-to-model comparison between two comparable multiple model of different complexity, [Tham88], the computer simulation of the multiple model based on a non-pulsatile hemodynamic submodel, described here, ran fifty times faster than its pulsatile counterpart. This advantage is significant when the multiple model is implemented on the AT class of personnal computers. The simulation ran faster than real-time.

The composition of the large-scale multiple model can further be extended by incorporating interacting and parallel gas transport submodels such as carbon dioxide and oxygen distribution which control the chemoreceptor reflexes, and nitrous oxide which causes a second gas effect. Addition of chemoreflex control was attempted by Fukui [Fuku81b] on an analog computer. With the development of the non-pulsatile anesthesia model, digital versions of such extensions are feasible.

A large-scale multiple model has been instrumental in analytically integrating some of the large amount of available information describing non-linear physiological beha-vior during anesthesia, and certainly more can be added. The framework for adding knowledge is available through the multiple model concept.

REFERENCES

[Allo76]   Allot, P.R., Steward, A. and Mapleson, W.W. 1976. Pharmacokinetics of halothane in the dog. Brit. J. Anaesth. 48:279.

[Ashm70]   Ashman, M.N., Blesser, W.B. and Epstein, R.M. 1970. A nonlinear model for the uptake and distribution of halothane in man. Anesthesiol. 33:419-428.

[Bedu79]   Beduhn, D.L. 1979. A model of anesthetic uptake and distribution. Masters Thesis. UW-Madison.

[Bene79]   Beneken, J.E.W. and Rideout, V.C. 1968. The use of multiple models in cardiovascular system studies: transport and perturbation methods. IEEE Trans BME 15:281-289.

[Dedr68]   Dedrick, R.L. and Bischoff, K.B. 1968. Pharmacokinetics in applications of the artificial kidney. Chem. Eng. Progr. Symp. 81(64):32-44.

[Dick68]   Dick, D.E. 1968. A hybrid computer study of major transients in the canine cardiovascular system. Ph. D. Thesis. University of Wisconsin-Madison.

[Fuku73]   Fukui, Y. 1972. A study of the cardiovascular-respiratory system using hybrid computer modelling. Ph. D. Thesis. University of Wisconsin-Madison.

[Fuku81a] Fukui, Y. and Smith, T.N. 1981. Interaction among ventilation, circulation, and the uptake and distribution of halothane - Use of a hybrid computer multiple model: I. The basic model. Anesthesiol. 54:107-118.

[Fuku81b] Fukui, Y. and Smith, T.N. 1981. Interaction among ventilation, circulation, and the uptake and distribution of halothane - Use of a hybrid computer multiple model: II. Spontaneous vs. controlled ventilation, and the effects of Carbon Dioxide. Anesthesiol. 54:119-124.

[Guyt81]   Guyton, A.C. 1981. Textbook of physiology. Saunders. Philadelphia.

[Jacq72]   Jacquez, J.A. 1972. Computational analysis in biology and medicine: Kinetics of Tracer-labeled Materials. Elsevier. New Tork.

[Kato80]   Katona, P.G. 1980. Automated control of physiological variables and clinical therapy. CRC Crit. Rev. Biomed. in Eng. 8(4):281-310.

[Kezd67]   Kezdi, P (Ed). 1967. Baroreceptors and Hypertension. Pergamon Press. NewYork.

[Levy79]   Levy, M.N. and Martin, P. 1979. Neural Control of the Heart. In "Handbook of Physiology. Section 2: The Cardiovascular System. Vol. 1: The Heart". Am. Physiol. Soc. Bethesda. Maryland.

[Lync81]   Lynch, C., Vogel, S. and Sperelakis, N. 1981. Halothane depression of myocardial slow action potentials. Anesthesiol. 55:360-368.

[Mapl73]   Mapleson, W.W. 1973. Circulation-time models of the uptake of inhaled aneaesthetics and data for quantifying them. Brit. J. Anaesth. 45:319.

[Mart70]   Martin, P. 1970. Automatica. 6:175.

[Midd72]   Middleman, S. 1972. (Ed). Transport phenomenon in cardiovascular system. John Wiley. New York.

[Mitc86]   Mitchell & and Gauthier Associates. 1986. Advanced continuous simulation language. Mitchell & Gauthier Associates. Concord, MA. 01742.

[Ride75]   Rideout, V.C. 1975. Mass Transport simulation using compartments and multiple modeling. ISA Trans. 14:109.

[Ride83]   Rideout, V.C. 1983. Linear analysis of the cardiovascular system. In "Integrated approaches to monitoring." Gravenstein, J.S. *et al* (ed). Butterworths.

[Ride85]   Rideout, V.C. and Tham, Q.Y.R. 1985. A non-pulsatile pressure-flow cardiovascular model. J. Clin. Monitoring. 1(1):90.

[Schi83]   Schils, G.F. 1983. A study of servo-anesthesia. Doctoral Thesis. UW-Madison.

[Schi87]   Schils, F.J., Sasse, F.J., Rideout, V.C. 1987. Automatic Control of Anesthesia Using Two Feedback Variables. Annals of Biomed Engr. 15:19-34.

[Seag85]   Seagard, J.L., Bosnjak, Z.J., Hopp, F.A. (Jr), Kotrly, K.J., Ebert, T.J. and Kampine, J.P. 1985. Chapter 12 of Effects of Anesthesia. Covino, B.G., Fozzard, H.A., Rehder, K. and Strichartz, G. (eds). Am. Physiol. Soc. Williams & Wilkins. Baltimore.

[Seag83]   Seagard, J.L., Hopp, F.A., Bosnjak, Z.J., Elegbe, E.O. and Kampine, J.P. 1983. Extent and Mechanism of halothane sensitization of the carotid sinus baroreceptors. 58:432-437.

[Seag82]   Seagard, J.L., Hopp, F.A., Donegan, J.H., Kalblleisch, J.H. and Kampine, J.P. 1982. Halothane and the carotid sinus reflex: Evidence for multiple sites of action. Anesthesiol. 57:191-202.

[Suna82]   Sunagawa, K., and Sagawa, K. 1982. Models of ventricular contraction based on time-varying elastance. CRC Crit. Rev. in Biomed. Eng. pp193-228.

[Tham88]   Tham, R.Q.Y. 1988. A Study of the Effects of Halothane on the Canine Cardiovascular System and Baroreceptor Control. Ph. D. Thesis. University of Wisconsin-Madison.

[Zwar72]   Zwart, A., Smith, T.N. and Beneken, J.E.W. 1972. Multiple model approach to uptake and distribution of halothane: the use of an analog computer. Comp. & Biomed. 5:228-238.

Index